Advances in the Health, Behavior, and Physiology of Honeybees and Other Pollinators

Guest Editors

Giovanni Formato
Franco Mutinelli

MDPI

Basel • Beijing • Wuhan • Barcelona • Belgrade • Novi Sad • Cluj • Manchester

Guest Editors

Giovanni Formato
Apiculture Laboratory
Istituto Zooprofilattico
Sperimentale del Lazio e della
Toscana "M. Aleandri"
Rome
Italy

Franco Mutinelli
National Reference Centre for
Beekeeping
Istituto Zooprofilattico
Sperimentale delle Venezie
Legnaro
Italy

Editorial Office

MDPI AG
Grosspeteranlage 5
4052 Basel, Switzerland

This is a reprint of the Special Issue, published open access by the journal *Insects* (ISSN 2075-4450), freely accessible at: https://www.mdpi.com/journal/insects/special_issues/6IUC18KFM1.

For citation purposes, cite each article independently as indicated on the article page online and as indicated below:

Lastname, A.A.; Lastname, B.B. Article Title. *Journal Name* **Year**, *Volume Number*, Page Range.

ISBN 978-3-7258-7190-2 (Hbk)
ISBN 978-3-7258-7191-9 (PDF)
https://doi.org/10.3390/books978-3-7258-7191-9

Contents

About the Editors

Giovanni Formato

Giovanni Formato, a Doctor of Veterinary Medicine and an internationally recognized expert in bee diseases and apiculture management, has extensive experience in veterinary public health and the food safety of animal products. He holds two Master's degrees, one in the Food Safety of Animal Products, and one in Pathology and General Apidology. He further specialized in Control and Self-Control of Food Products at the Catholic University of the Sacred Heart and in Epidemiological Methods in Veterinary Public Health at the Istituto Zooprofilattico Sperimentale dell'Abruzzo e del Molise. Since 1998 he has been working at the Istituto Zooprofilattico Sperimentale del Lazio e della Toscana M. Aleandri (IZSLT), where he has headed the Laboratory of Apiculture since 2009. His work focuses on diagnostics, research, training, and international cooperation in support of sustainable beekeeping. He collaborates with governmental institutions, universities, research laboratories, and international organizations such as the Food and Agriculture Organization and the World Organization for Animal Health, serving since 2021 as the Contact Point for the WOAH Collaborating Centre for Good Beekeeping Management Practices and Biosecurity Measures in the Apiculture Sector. Since 2022 he has been a member of the DISCONTOOLS expert group on Varroosis. Dr. Formato coordinates the B-THENET (2023–2026) Horizon Europe Thematic Network on best practices and innovations for sustainable beekeeping. He is partner in INSIGNIA-EU, aimed at standardizing protocols for environmental pollution monitoring using honey bees, and in BETTER-B, where he led the work package on the future of EU beekeeping under global change pressures. He is a member of COLOSS, co-founder and President of the Scientific Society of Veterinarians in Apiculture (SVETAP), co-founder of the International Society of Veterinarians in Beekeeping (Vets4Bees), and member of the Bee Health Commission of APIMONDIA.

Franco Mutinelli

Franco Mutinelli, Doctor of Veterinary Medicine, directs the Functional Animal Experimental and Welfare Department (2013) and the SCS3, the Specialist Diagnostics, Histopathology and Sanitary Entomology (2003) of Istituto Zooprofilattico Sperimentale delle Venezie in Legnaro (Padova), Italy. He is also the Director of the National Reference Centre for Beekeeping (2003), the National Reference Laboratory for Bee Diseases (2018), the FAO Reference Centre for apiculture: health, biosecurity and AMR (2020) shared with Istituto Zooprofilattico Sperimentale del Lazio e Toscana (Rome) and the Head of the Regional Centre for Beekeeping in Veneto (1995). Since June 2021, he has been the Director of the National Reference Centre for Assisted Interventions with Animals. He graduated in Veterinary Medicine in 1983 from the University of Bologna, followed by the European College of Veterinary Pathologists in 1998, and, in 2006, the Executive Master in Management for Healthcare Companies (EMMAS V, Bocconi University, Milan). His expertise concerns animal pathology with specific competences in the prevention, diagnosis and control of honey bee diseases, the elaboration of surveillance programmes for Varroa destructor, *Aethina tumida* and Tropilaelaps, the quality and salubrity of hive products, the use of honey bees for environment monitoring, and legislation. He is part of several collaborations in national and international research projects. He participated as a speaker and tutor in the Better Training for Safer Food (BTSF) "Animal Health of Bees" (2010–2017), the Live Animals "Import Control of Bees and Bumblebees" (2017–2024) and the "Companion Animals Movement" (2017–2024) in TAIEX events (2018–2024) related to the protection and improvement of the health of bees and the production of the bee and the beehive. He is also a member of the EU Expert Group for Bee Diseases and member of the Executive Committee of Coloss Association.

insects

MDPI

Editorial

Advances in the Health, Behavior, and Physiology of Honeybees and Other Pollinators

Giovanni Formato [1] and Franco Mutinelli [2,*]

1 FAO RC for Apiculture: Health, Biosecurity and AMR, WOAH Collaborating Centre for Good Beekeeping Management Practices and Biosecurity Measures in the Apiculture Sector, Istituto Zooprofilattico Sperimentale del Lazio e della Toscana, Via Appia Nuova 1411, 00178 Rome, Italy; giovanni.formato@izslt.it

2 NRL for Honey Bee Health, NRL for Honey Bee Health and FAO RC for Apiculture: Health, Biosecurity and AMR, Istituto Zooprofilattico Sperimentale delle Venezie, Viale dell'Università 10, 35020 Legnaro, Italy

* Correspondence: fmutinelli@izsvenezie.it; Tel.: +39-049-8084287

The health, behavior, and physiology of honeybees and other pollinators are critical areas of research with profound implications for biodiversity, agriculture, and ecosystem stability. As keystone species, pollinators are essential for the reproduction of many plants, influencing both natural and agricultural ecosystems. Recent scientific advancements have highlighted the complex interplay of factors affecting pollinator health, including environmental stressors, pathogens, pesticides, and habitat loss. Understanding these dynamics is vital for developing strategies to mitigate their decline and ensure ecological and agricultural sustainability.

This Special Issue of *Insects* on "Advances in the Health, Behavior, and Physiology of Honeybees and Other Pollinators" aims to consolidate cutting-edge research in this pivotal field. Aligning with the journal's focus on the biology, ecology, and management of insects, this Special Issue will explore innovative findings and foster interdisciplinary discourse, and researchers are invited to submit original research articles, reviews, and case studies that contribute to our understanding of pollinator health and behavior, their physiological responses to various environmental pressures, and the development of conservation strategies.

Topics of interest include, but are not limited to, the impact of pesticides and pathogens, the effects of climate change, habitat requirements, genetic diversity, current and innovative management practices, and bee welfare. We look forward to receiving contributions that will drive forward the scientific understanding of and conservation efforts for these indispensable insects.

The Special Issue "Advances in the Health, Behavior, and Physiology of Honeybees and Other Pollinators" has published 10 papers, including 8 research articles, 2 review articles, and 1 communication article. This Editorial explores how beekeeping practices, parasite control measures, and exposure to toxic substances influence the health and survival of honey bee colonies.

Al-Ghzawi et al. [Contribution 1] evaluated the effects of preheat hardening on honeybee queen egg-laying and the flight performance of their daughter workers. Eggs from caged queens were incubated at either 41 °C (pre-heat-treated queens, pH-TQs, $n = 12$) or 34.5 °C (non-heat-treated queens, nH-TQs, $n = 12$) for 15 min. After mating, both groups were introduced into queenless hives. Colonies headed by pH-TQs reared significantly more brood and adult workers—especially in summer—collected more pollen, and produced fewer drone brood and swarm cells than nH-TQ colonies. These results suggest preheat hardening enhances reproductive potential and adaptability in honeybees,

 https://doi.org/10.3390/insects17020231

offering a practical approach for improving colony performance and resilience in the face of global honeybee declines.

Self-regulatory foraging in honey bees (*Apis mellifera*) was examined by Jones et al. [Contribution 2] using Perceptual Control Theory (PCT). We developed a PCT-based model to describe how bees maintain goal-directed behavior toward a sucrose-rich feeding site while overcoming wind disturbances. In controlled experiments, 13 out of 14 bees successfully adjusted flight paths to reach the target, showing substantial individual variability in strategies. Ultimately, the bees consistently adopted a headwind approach rather than tailwind or crosswind. These findings support the view that honey bee foraging is self-regulative—dynamic, non-linear, and goal-directed—rather than a simple input–output sequence. PCT provides a framework to model such teleonomic behaviors, emphasizing functional purpose in behavioral regulation.

The increasing demand for honey bee (*Apis mellifera*) pollination has intensified efforts to preserve genetic diversity and support breeding programs. Cryopreservation of drone semen offers a solution without the risks associated with live animal transport. Egyptien et al. [Contribution 3] evaluated a one-step, antibiotic-free slow-freezing protocol under field conditions using in vitro and in vivo assessments. Post-thaw semen viability decreased by 37%, yet virgin queens inseminated with frozen–thawed semen produced brood at rates comparable to those inseminated with fresh semen, although winter survival was higher in the fresh semen group. These findings indicate that frozen–thawed semen can support viable queen insemination and brood production, highlighting the potential for cryobanking while emphasizing the need for protocol refinement to enhance semen quality and queen longevity.

Honeybee combs are central to colony development, serving as sites for brood rearing, food storage, and other essential activities. As combs age through successive brood cycles, their structural and physicochemical properties deteriorate, negatively affecting colony health and productivity. This review by Meng et al. [Contribution 4] synthesizes current knowledge on comb functions, aging processes, and associated impacts, including altered cell dimensions, smaller worker bees, reduced foraging efficiency, and compromised honey and beeswax quality. Honeybees mitigate these effects through hygienic behaviors, comb rebuilding, and enhanced immune and detoxification responses. Systematic replacement of old brood combs is emphasized as a key management practice to maintain colony strength, optimize bee health, and support sustainable apiculture.

Honey bees (*Apis mellifera*) are globally important pollinators and economically valuable insects, yet they face numerous stressors that threaten colony health. Adequate nutrition is critical for maintaining strong, resilient colonies, requiring balanced intake of proteins, carbohydrates, lipids, amino acids, vitamins, minerals, water, and essential sterols, with multi-floral pollen being ideal. Colonies in low-diversity agroecosystems often experience reduced brood rearing and longevity, increasing susceptibility to parasites and diseases. Nutritional supplements have been developed to support colony health, but nutrient deficiencies may persist. In this review, Ansaloni et al. [Contribution 5] highlight the relationship between dietary requirements and colony well-being, emphasizing the need for supplements that better mimic natural diets and future research to evaluate long-term impacts on colony resilience, particularly in resource-limited environments.

Honey bee health and productivity are strongly shaped by management and biosecurity practices. Gratzer et al. [Contribution 6] analyzed 744 records from 191 field studies (1995–present), categorizing interventions as good beekeeping practices (17.2%) or biosecurity measures (82.8%) across management themes, pathogens, regions, and seasons. Most studies focused on *Apis mellifera* in Europe (34.6%) and North America (33.4%). Varroa control dominated (57.0%), primarily through using soft acaricides (58.5%), while general

 https://doi.org/10.3390/insects17020231

colony management (17.2%) and American foulbrood interventions (antibiotics 41.7%) were also common. Seasonal trends showed most interventions occurred from August to October, with foulbrood peaks in spring–early summer. This dataset highlights regional research gaps and provides a structured framework to guide evidence-based decision-making in beekeeping.

Varroa destructor is a key driver of global honey bee colony declines, and miticide applications are widely used to control infestations. While forager exposure has been studied extensively, nurse bees are also vulnerable due to direct contact and residual hive exposure. Kim et al. [Contribution 7] assessed the toxicities of single and consecutive applications of three synthetic miticides (fluvalinate, coumaphos, amitraz) and two organic miticides (formic acid, oxalic acid) at field-realistic doses. Synthetic miticides caused minimal mortality, whereas organic miticides—particularly formic acid—significantly reduced survivorship. Consecutive treatments largely mirrored the toxicity of the organic agent alone, with the fluvalinate–formic acid combination showing no significant adverse effects. These results emphasize careful miticide use and inform safer strategies to protect in-hive nurse bees.

The small hive beetle (*Aethina tumida*) threatens honeybee (*Apis mellifera*) populations in Europe, with potential spread from Calabria, Italy. Early detection is critical, as current methods often fail, leading to colony losses. Ghisellini et al. [Contribution 8] developed a biosensor to detect *Kodamaea ohmeri*, a yeast symbiotically associated with *A. tumida*, in honey samples. Using a quartz crystal microbalance with a gold-plated surface functionalized with a specific antibody, the biosensor identified peptides linked to *K. ohmeri*, offering a potential early-warning system for infestations. The preliminary results demonstrate the feasibility of this approach to support rapid and targeted colony management. Further miniaturization, cost reduction, and beekeeper training are recommended to enhance adoption and improve honeybee health and productivity.

Lambda-cyhalothrin (LCY), a widely used pyrethroid insecticide, is toxic to honey bees, yet its molecular effects during early development are poorly understood. Vasamsetti et al. [Contribution 9] examined chronic sublethal LCY exposure in larvae and adults, comparing solvent-treated and LCY-treated groups. Differential gene expression analysis revealed 1128 altered genes in larvae and 168 in adults, with stage-specific effects on 125 larval and 25 adult genes. LCY disrupted cuticle formation, sulfur metabolism, oxidoreductase activity, and neuropeptide signaling in larvae, while affecting redox balance, peptide receptor signaling, and monoamine transport in adults. Both stages showed neuroactive signaling disruptions, with larvae additionally showing impaired motor function, amino acid metabolism, and glycolysis. Downregulation of chitin metabolism and antioxidant genes in larvae suggests weakened exoskeletal integrity. These findings highlight lasting molecular consequences of early-life LCY exposure and underscore the need for safer pesticide practices to protect pollinators.

Honey bees (*Apis mellifera*) are critical pollinators, yet pesticide exposure poses risks to their health. Hu et al. [Contribution 10] investigated the sublethal effects of atrazine on honey bee sucrose sensitivity and the underlying molecular mechanisms via transcriptomic analysis. Atrazine exposure significantly reduced sucrose sensitivity, likely by suppressing brain genes involved in cognition. Differentially expressed genes were associated with neurodegenerative processes and behavioral regulation, indicating that atrazine can disrupt neural function and foraging behavior. These findings provide new insights into the neurophysiological and behavioral impacts of atrazine, highlighting potential risks to pollinator health and informing strategies for safer pesticide use and environmental regulation.

 https://doi.org/10.3390/insects17020231

This Special Issue, "Advances in the Health, Behavior, and Physiology of Honeybees and Other Pollinators", brings together a diverse and relevant collection of contributions addressing key challenges and emerging knowledge in honey bee science. The ten papers published in this Special Issue, including original research articles, review articles, and a communication, highlight the interdisciplinary nature of current research in apiculture and pollinator biology.

Pollinators continue to face unprecedented pressures from pathogens, environmental change, habitat loss, and anthropogenic stressors. At the same time, advances in physiology, behavior, diagnostics, biosecurity measures, and management practices are expanding our ability to understand and protect these essential animals. The studies presented in this Special Issue reflect both scientific rigor and practical relevance, bridging fundamental research with applied perspectives that are critical for sustainable beekeeping and ecosystem resilience.

We hope that this compilation will stimulate further research, foster international collaboration, and contribute to evidence-based strategies for protecting honeybees and other pollinators. Continued investment in pollinator science is not only a scientific priority but also a societal necessity, given the central role of pollinators in food security, biodiversity, and ecosystem services.

We trust that readers will find inspiration in these contributions and that this collection will serve as a foundation for future advances in pollinator research.

Funding: This research received no external funding.

Institutional Review Board Statement: Not applicable.

Data Availability Statement: Data can be requested to the authors of the contributions collected in this SI.

Acknowledgments: We are grateful to all authors who contributed their work, sharing innovative findings and thoughtful analyses. We also extend our sincere thanks to the reviewers for their careful evaluations and constructive feedback, which ensured the scientific quality of the published articles. Their collective efforts have made this Special Issue a meaningful resource for researchers, practitioners, and policymakers engaged in pollinator health and conservation. We thank the *Insects* Editorial Office for their professional support throughout the development of this Special Issue and for facilitating the dissemination of this body of work to the broader scientific community.

Conflicts of Interest: The authors declare no conflicts of interest.

List of Contributions:

1. Al-Ghzawi, A.A.; Zaitoun, S.T.; Al-Sabi, M.N.S.; Mazari, S.S.; Al-Omari, I.M.; Altalhi, M.S. Assessing the Productivity of Colonies Headed by Preheated Honeybee Queens. *Insects* **2025**, *16*, 858. https://doi.org/10.3390/insects16080858.
2. Jones, I.T.; Grice, J.W.; Abramson, C.I. Living Control Systems: Exploring a Teleonomic Account of Behavior in *Apis mellifera*. *Insects* **2025**, *16*, 848. https://doi.org/10.3390/insects16080848.
3. Egyptien, S.; Ponthier, J.; Ectors, F.; Thibaut, B.; Deleuze, S. Evaluation of a Simple Antibiotic-Free Cryopreservation Protocol for Drone Semen. *Insects* **2025**, *16*, 50. https://doi.org/10.3390/insects16010050.
4. Meng, Q.; Huang, R.; Yang, S.; Jiang, W.; Tian, Y.; Dong, K. An Overview of the Adverse Impacts of Old Combs on Honeybee Colonies and Recommended Beekeeping Management Strategies. *Insects* **2025**, *16*, 351. https://doi.org/10.3390/insects16040351.
5. Ansaloni, L.S.; Kristl, J.; Domingues, C.E.C.; Gregorc, A. An Overview of the Nutritional Requirements of Honey Bees (*Apis mellifera* Linnaeus, 1758). *Insects* **2025**, *16*, 97. https://doi.org/10.3390/insects16010097.

6. Gratzer, K.; Musalkova, V.; Brodschneider, R. How Science Supports Honey Bees: Identification of Research on Best Practices in Beekeeping. *Insects* **2025**, *16*, 1025. https://doi.org/10.3390/insects16101025.
7. Kim, H.; You, E.; Cha J.; Lee, H.S.; Kim, Y.H. Toxicity of Consecutive Treatments Combining Synthetic and Organic Miticides to Nurse Bees of *Apis mellifera*. *Insects* **2025**, *16*, 657. https://doi.org/10.3390/insects16070657.
8. Ghisellini, P.; Garbati, P.; Pietropaoli, M.; Cersini, A.; Pietrella, G.; Rando, C.; Giacomelli, L.; Ottoboni, S.; Formato, G.; Eggenhöffner, R. A Novel Biosensor for the Early Detection of *Aethina tumida* via *Kodamaea ohmeri* in Honeybee Colonies. *Insects* **2025**, *16*, 363. https://doi.org/10.3390/insects16040363.
9. Vasamsetti, B.M.K.; Chon, K.; Kim, J.; Choi, M.; Kim, B.-S.; Yoon, C.-Y.; Hwang, S.; Park, Y.-H. Chronic Larval Exposure to Lambda-Cyhalothrin Alters Gene Expression in Both Larval and Adult Honey Bees (*Apis mellifera*). *Insects* **2025**, *16*, 833. https://doi.org/10.3390/insects16080833.
10. Hu, X.; Xu, Z.; Xu, J.; Ma, G.; Pan, Y.; Cai, M.; Lin, Z.; Ji, T.; Wang, K. Impact of Atrazine on Sucrose Sensitivity in Honey Bees. *Insects* **2025**, *16*, 491. https://doi.org/10.3390/insects16050491.

https://doi.org/10.3390/insects17020231

insects

MDPI

Article

Assessing the Productivity of Colonies Headed by Preheated Honeybee Queens

Abd Al-Majeed Al-Ghzawi [1,*], Shahera Talat Zaitoun [2], Mohammad Nafi Solaiman Al-Sabi [3], Salem Saleh Mazari [4], Ilham Mustafa Al-Omari [1] and Maqbool Saed Altalhi [5]

1 Department of Plant Production, Faculty of Agriculture, Jordan University of Science and Technology, P.O. Box 3030, Irbid 22110, Jordan; emamassad@gmail.com
2 Department of Plant Production & Protection, Faculty of Agricultural Technology, Al-Balqa Applied University, P.O. Box 206, Salt 19117, Jordan; zaitoun@bau.edu.jo
3 Department of Basic Medical Veterinary Sciences, Jordan University of Science and Technology, P.O. Box 3030, Irbid 22110, Jordan; mnalsabi@just.edu.jo
4 Ministry of Agriculture, P.O. Box 2099, Amman 11181, Jordan; ssmazarijo@outlook.com
5 Taif Beekeepers Cooperative Society, TIDA2375, Taif 26724, Saudi Arabia; maqbool81@gmail.com
* Correspondence: ghzawi@just.edu.jo; Tel.: +96-27-9528-9605

Simple Summary: Preheat hardening in the immature stages causes marked changes in the morphology of honeybee workers. In this study, we investigated the effects of such preheat hardening on the reproductive capacity of honeybee queens and the subsequent productivity of the workers. We found that heat-treated queens produced workers that had significantly increased numbers of brood cells and worker adults, especially during the hot summer months. These workers also managed to collect and store more pollen, had less tendency to rear drone brood, and constructed fewer swarm cells than non-heat-treated workers. We therefore propose the use of heat hardening to compensate for the current loss of honeybees globally.

Abstract: This study investigated the effects of preheat hardening on the egg-laying capacity of honeybee queens and the flight performance of their daughter workers. A honeybee queen was confined in a cage with a two-section frame for 12 h. Then, 48 h old eggs from one section were incubated for 15 min at 41 °C and 70% relative humidity (RH). The queens (n = 12) raised in this section were named the pre-heat-treated queens (pH-TQs). Eggs from the second section were exposed to 34.5 °C and 70% RH for 15 min, and the queens raised in this section were named the non-heat-treated queens (nH-TQs) (n = 12). After mating, both groups were introduced to queenless hives in order to produce workers. The results show that, during the study period (2021 and 2022), the colonies headed by the pH-TQs reared significantly more brood cells and worker adults (especially during the summer), collected and stored more pollen, had less tendency to rear drone brood, and constructed fewer swarm cells than the colonies headed by the nH-TQs. Preheat hardening can be a promising method for improving honeybee workers' reproductive potential and adaptability, allowing for better adaptation to environmental changes, thus compensating for the mass fatalities of honeybees globally.

Keywords: beehives productivity; flight performance; honeybee behavior; preheat hardening; climate change effects; environmental stressors

1. Introduction

In recent decades, heat has been increasingly used for the control of various pests causing economic harm, as well as for the control of food and non-food items in stores [1–3], plant diseases [4], human therapy [5], Varroa mites inside hives [6,7], certain honeybee fungal diseases [8], and viral diseases [9]. In a more advanced step of utilizing heat, several studies have demonstrated that the thermal manipulation of honeybee queens with non-lethal heat shock during the egg stage has beneficial effects on the subsequent physiology and behavior of their daughter workers. This is defined as rapid heat hardening and has been practiced to improve the survival of bees [10,11]. Honeybees show a marked response to heat exposure in terms of physiology [12], including changes in antioxidant and detoxification enzymes [13]; adjustments in metabolism and nutrient supply [14,15]; an increase in gene expression levels when exposed to heat, cold, and starvation stress [16–18]; and, in queens and workers, even a change in color [19]. In terms of the behavioral responses to heat exposure [20,21], honeybees also show changes in their development [22], immunity [23], phenology [24], ability to learn [20,25], behavioral performance [26,27], disease susceptibility [28], and flight behavior [21,29].

To overcome temperature fluctuations and prevent extremely high temperature levels inside the hive, honeybee workers in the colony strictly keep it within a range of 33–36 °C. This is because the brood is very sensitive to changes in the hive temperature [30]. It has been previously reported that the danger of overheating is mitigated by behavioral changes in the workers, including the activation of fanning behavior, i.e., rapid vibration of the wings [31]; collecting and depositing water inside the hive to allow for evaporative cooling [32]; relocating bees outside the entrance to facilitate continuous ventilation and evaporation inside the hive [33]; and shielding honey and brood combs from sources of external heat [34]. These and other behaviors have allowed the honeybee subspecies to successfully live in a wide range of climates, from hot and dry to tropical and temperate climates.

To the best of our knowledge, the effects of the heat shock treatment of honeybee queens during the developmental stages on brood rearing and worker populations have not yet been investigated. We hypothesized that the exposure of honeybee queens to non-lethal heat shock during the egg stage could improve their viability and workers' honey production. The results of this study help determine whether exposure to high temperatures during the developmental stages adjusts the queen's egg-laying capacity. Furthermore, they provide a scientific basis for further exploration of the temperature adaptation mechanism of queen honeybees, ultimately helping to formulate a comprehensive improvement program.

2. Materials and Methods

2.1. Queen Rearing and Thermal Manipulation

The origin of the experimental colonies (n = 24) was healthy honeybees headed by the local race, the Syrian honeybee (*Apis mellifera* syriaca), selected from a privately owned apiary in northern Jordan. The colony queens were reared in April 2020, which is known to be the prime month for open mating in the Jordan Valley [35]. A queen from a colony was confined in a queen excluder with a two-section frame for 12 h to produce eggs of approximately the same age. Just before egg hatching (48 h old), both sections were taken from the colony after brushing the nurses off, and they were transferred to the lab at Agri-JUST. The first section was incubated at 41 °C and 70% relative humidity (RH) for 15 min and then returned to its mother hive. After egg hatching, the first-instar worker larvae were grafted to queen cups to be raised in a queenless hive, following standard apicultural procedures [36]. The queens raised under this treatment were named the pre-heat-treated queens (pH-TQs). The second section was exposed to 34.5 °C and 70% RH for 15 min and then returned to its mother hive. After egg hatching, the same procedure was followed to rear non-heat-treated queens (nH-TQs) as the control. Just before pupal emergence, plastic mesh cages were fitted over the occupied pupal queen cells of both groups to retain the emerging virgin queens. The emerging virgin queens in both groups were removed from the plastic mesh cages and introduced to queenless mating mini-hives. These hives were inspected two weeks later to ensure the success of the introduction and the onset of oviposition. From each treatment group, 12 queens were introduced to queenless experimental colonies following standard procedures [36]. Six months later, the experimental colonies were well established, occupying 8–10 frames. Before the commencement of the study in November, each colony was treated with amitraz against Varroa mites, and all experimental colonies were equalized to include a similar amount/number of stored food, brood, and adult bees.

During the investigation, the colonies were subjected to various apiculture practices, including shading, feeding, supering, and external pest control, as needed. The same procedure was repeated to produce new colonies headed by new queens treated in the same manner as above for use in the second year of the study.

2.2. Apicultural Practices

As usually practiced by beekeepers in Jordan, from November to April, the experimental colonies were placed on a designated farm in Shuna North in the Jordan Valley (32°37″ N, 35°36″ E), which is 200 m below sea level. Throughout this period, warm weather conditions prevailed in the Jordan Valley, together with the early flowering of cultivated and wild plants. As the temperatures increased and the wild and cultivated plants dried out in the Jordan Valley, the experimental colonies were relocated to the highlands in an apiary located in the Karak district (31°16″ N, 35°28″ E), with an elevation of 920 m above sea level. The same practices were repeated in the second year of the study. To reduce bee drifting, the locations of the hives were selected to ensure good protection from the wind. A total of 24 hives headed by pH-TQs and nH-TQs (n = 12 each) were randomly placed in each location in two rows; the rows were spaced four meters apart, while the hives in each row were spaced two meters apart.

2.3. Measurement of Worker and Drone Brood Areas

To estimate the cumulative worker and drone production and emergence patterns in a colony, the sealed brood area of each caste was measured throughout the season. This was carried out by placing a clear plastic grid marked out in 5 × 5 cm squares over the brood on one side of each frame and estimating the number of sealed cells in each quadrat [37].

2.4. Count of Adult Workers

The numbers of worker bees in the pH-TQ and nH-TQ colonies were counted at 21-day intervals throughout the experiment using the procedures described by Gerig [37]. The same counting method was employed in visual estimations of the numbers of brood, drone brood, and stored pollen cells and adult workers per unit area. When two counts were recorded in the same month, the average was calculated.

2.5. Count of Swarm Queen Cells

All pH-TQ and nH-TQ colonies were inspected at 10-day intervals throughout the swarming period. Queen cups and swarm queen cells with laid eggs or larvae in brood chambers were counted and destroyed. This was carried out between February and May during the study period from 2021 to 2022.

2.6. Reporting Pollen Foraging Percentage per Hour

The number of foragers returning to the entrance of each experimental colony was counted for a one-minute duration on an hourly basis between 06.00 AM and 4.00 PM for one day every two weeks. Pollen foragers were defined as workers returning to their colonies with a pollen load, while non-pollen foragers were defined as workers returning with no pollen load [38]. The pollen foraging percentage of a colony was calculated by dividing the number of returning bees with a pollen load by the total number of returning worker foragers to their hives and multiplying by 100.

2.7. Count of Stored Pollen Cells

The amount of pollen stored inside the colonies was measured using the same procedure used for bee and brood counting [37].

2.8. Weight of Pollen Loads

Pollen samples were collected using pollen traps fitted to the entrances of both the pH-TQ and nH-TQ colonies for one day every two weeks. The accumulated pollen was removed, cleaned, and dried at room temperature. Ten pollen loads from each hive were randomly selected from each pollen trap on each sampling date. The weight of each pollen load was obtained to the nearest 0.1 milligram (mg) using a technical balance (Scientech SP 350, Scientech, Boulder, CO, USA).

2.9. Statistical Analysis

All statistical analyses were performed using the R statistical system, Version 4.4.3 [39]. Data are expressed as means ± standard deviation (SD). A repeated-measures ANOVA with a generalized eta-squared test and the F statistic was employed to compare different parameters in all treatment groups. Each treatment was replicated twelve times. Differences were considered significant at $p < 0.05$.

3. Results

3.1. Brood-Rearing Activities

The colonies headed by the pH-TQs and nH-TQs showed a triple cycle of brood cell numbers within the two study years of 2021 and 2022 (Figure 1). Both colony groups started brood rearing based on the availability of nectar and pollen plants and the suitability of weather conditions in the Jordan Valley, reaching the entire season's maximum peak in March (35,700 and 27,512 cells, respectively). The pH-TQ colonies reared significantly more brood areas ($p < 0.001$) than the nH-TQ colonies during the steep rise in brood rearing between February and March. After the colonies migrated to the highlands, where the air temperature and wildflowers were suitable, the pH-TQ colonies showed a significantly higher ($p < 0.001$) second brood peak than the nH-TQ colonies in May (28,454 and 22,454 cells, respectively). The critical period for beekeeping started with the increase in air temperature and the drying of wildflowers by June, and, thereafter, both colony groups showed a decrease in brood-rearing activities. The pH-TQ colonies were less affected by high-temperature stress, and they reared significantly larger ($p < 0.001$) brood areas than the nH-TQ colonies from May to the end of the season. In the autumn, when the air temperature was suitable and certain fruit trees began to flower, the pH-TQ colonies showed a third brood peak with a significantly larger ($p < 0.001$) brood area than the nH-TQ colonies (20,191 and 10,891 cells, respectively). A repeated-measures ANOVA showed a significant difference in the number of brood cells between the pH-MQ and nH-MQ groups ($F(1, 574) = 147.10$, $p < 0.001$).

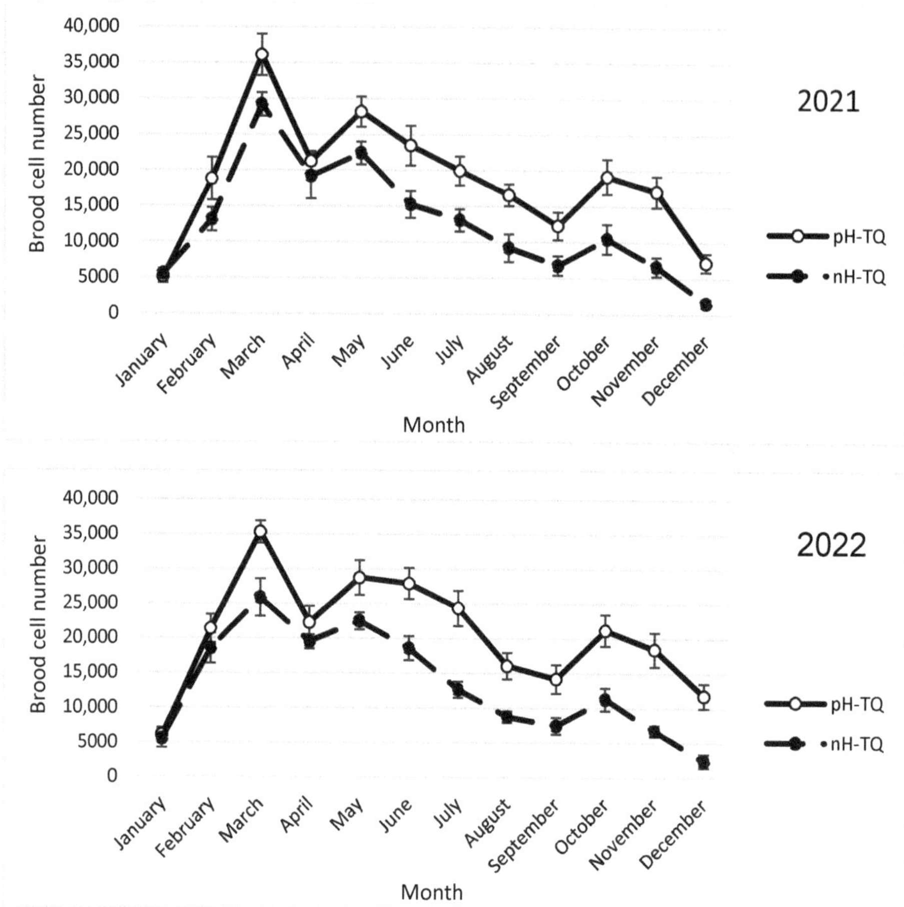

Figure 1. Number of worker brood cells (±standard deviation) of heat-treated honeybee queens (nH-TQ) and thermally manipulated honeybee queens in their immature stages (pH-TQ) in two consecutive years, 2021 and 2022.

3.2. Honeybee Worker Population

The annual cycle of the bees corresponded to the rate of brood rearing, and the maximum number of adults lagged behind the maximum rate of brood rearing. Both groups (the pH-TQ and nH-TQ colonies) showed a triple cycle of the worker population during the two study years of 2021 and 2022 (Figure 2). The pH-TQ colonies produced a significantly larger ($p < 0.001$) worker population than the nH-TQ colonies in April (35,033 and 28,366 workers, respectively) and in June (31,466 and 25,866 workers, respectively). During the summer, the pH-TQ colonies showed a steep reduction in worker numbers, while the nH-TQ colonies showed a sharp reduction in their worker numbers. The pH-TQ colonies showed a higher heat tolerance during the summer months and maintained significantly higher ($p < 0.001$) worker numbers throughout the period from June to the end of the year. After the cessation of the hot summer months, both groups of honeybee colonies, headed by the pH-TQs and nH-TQs, showed a triple peak in the worker population, with a significantly larger ($p < 0.001$) population in the pH-TQ colonies than in the nH-TQ colonies (22,625 and 16,600 cells, respectively). The seasonal fluctuations in brood-rearing activities and worker bee populations were fairly constant in terms of both time and relative magnitude throughout the two years of the study. The brood-rearing performance and population dynamics of the adult bees tended to follow the same general cycles in the second year of the study (Figure 2). A repeated-measures ANOVA showed a significant difference in the monthly changes in worker numbers between the pH-MQ and nH-MQ groups (F (1, 574) = 47.68, $p < 0.001$).

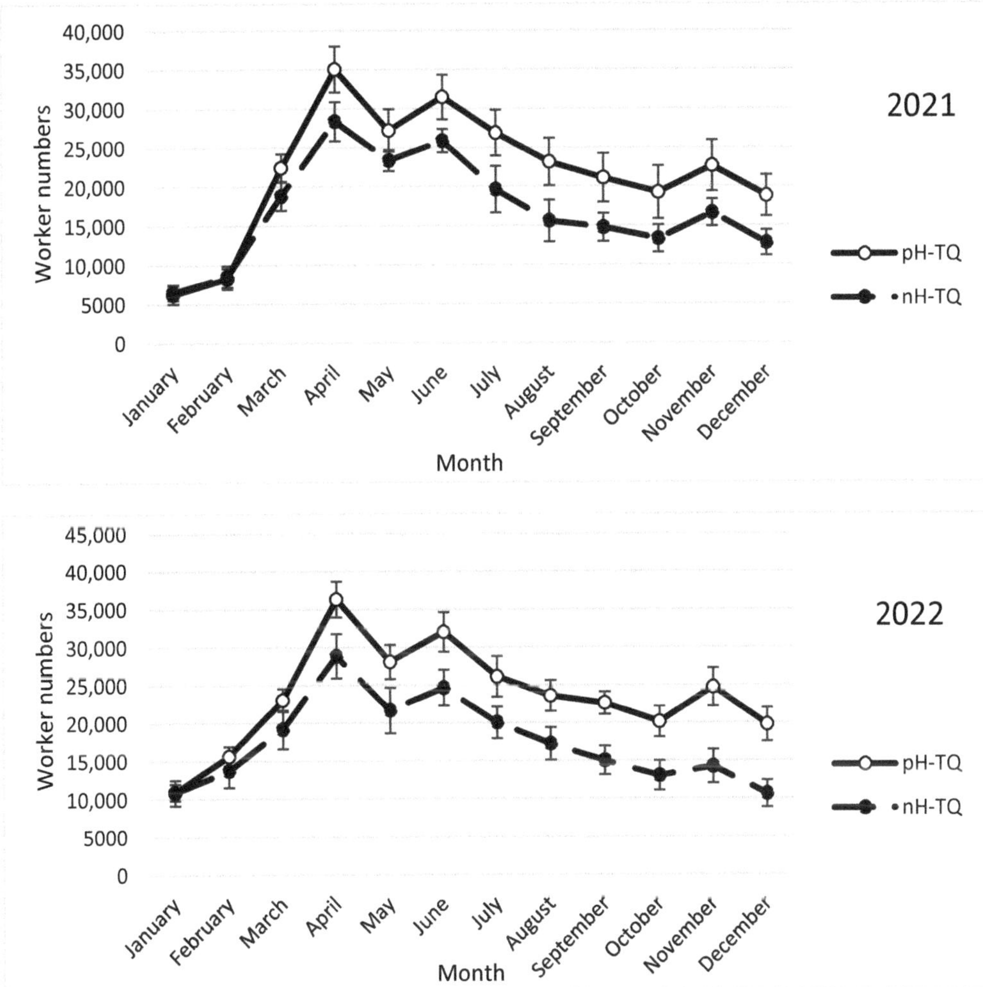

Figure 2. Number of honeybee workers (±standard deviation) of non-heat-treated honeybee queens (nH-TQ) and thermally manipulated honeybee queens in their immature stages (pH-TQ) in two consecutive years: 2021 and 2022.

3.3. Drone Brood Rearing

The colonies headed by the pH-TQ and nH-TQ queens showed a single cycle of drone brood cell numbers within the two study years of 2021 and 2022 (Figure 3). In both groups of colonies, drone brood rearing began in February and ceased in June. The nH-TQ colonies reared significantly more ($p < 0.001$) drone brood cells (310.3 cells) than the pH-TQ colonies (1675.0 cells in 2021). A repeated-measures ANOVA showed a significant difference in the number of drone brood cells between the pH-MQ and nH-MQ groups (F (1, 574) = 102.13, $p < 0.001$).

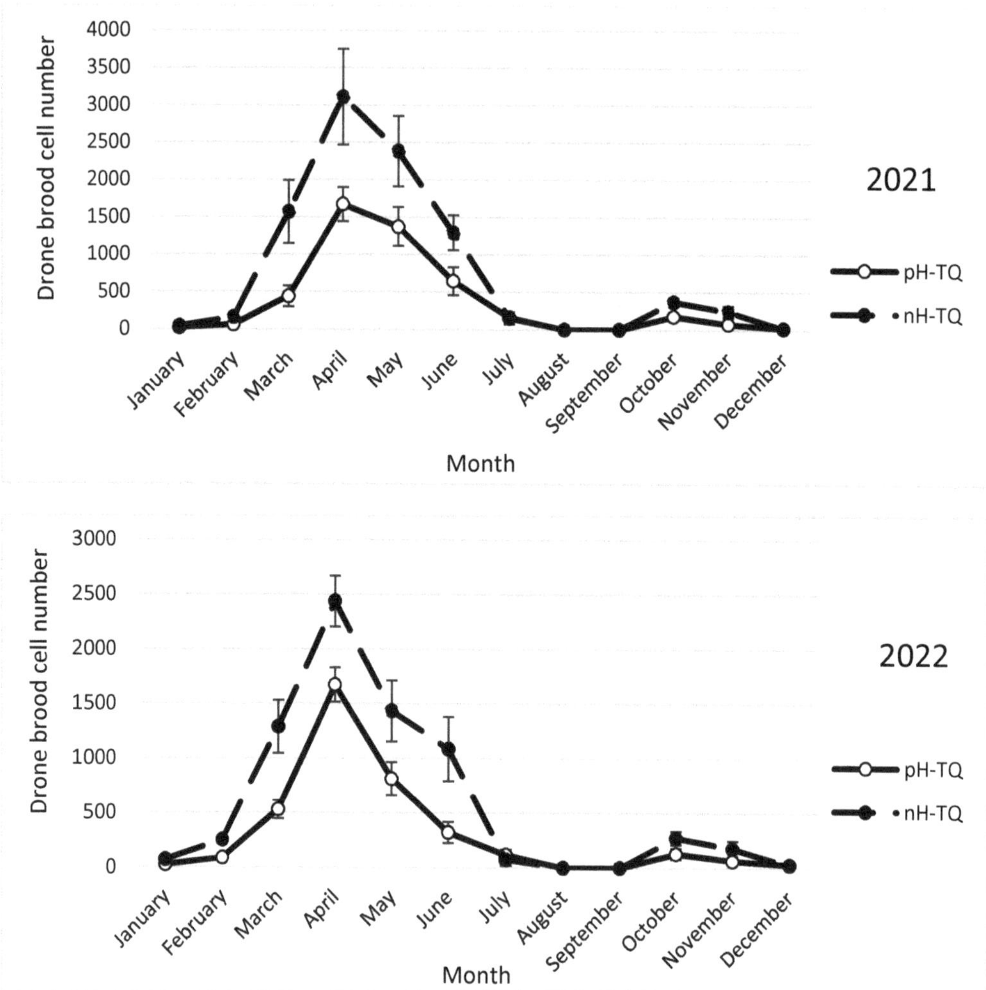

Figure 3. Drone brood cell numbers (±standard deviation) of non-heat-treated honeybee queens (nH-TQ) and thermally manipulated honeybee queens in their immature stages (pH-TQ) in two consecutive years: 2021 and 2022.

3.4. Swarm Cells

The pH-TQ and nH-TQ colonies showed swarm cell construction between March and May of each year of the study period (Figure 4). In the first year of the investigation, the nH-TQ colonies constructed significantly ($p < 0.001$) more swarm cells in April (120 cells) than the pH-TQ colonies (74.8 cells). A repeated-measures ANOVA showed a significant difference in the number of swarm queen cells between the pH-MQ and nH-MQ groups (F (1, 574) = 139.81, $p < 0.001$).

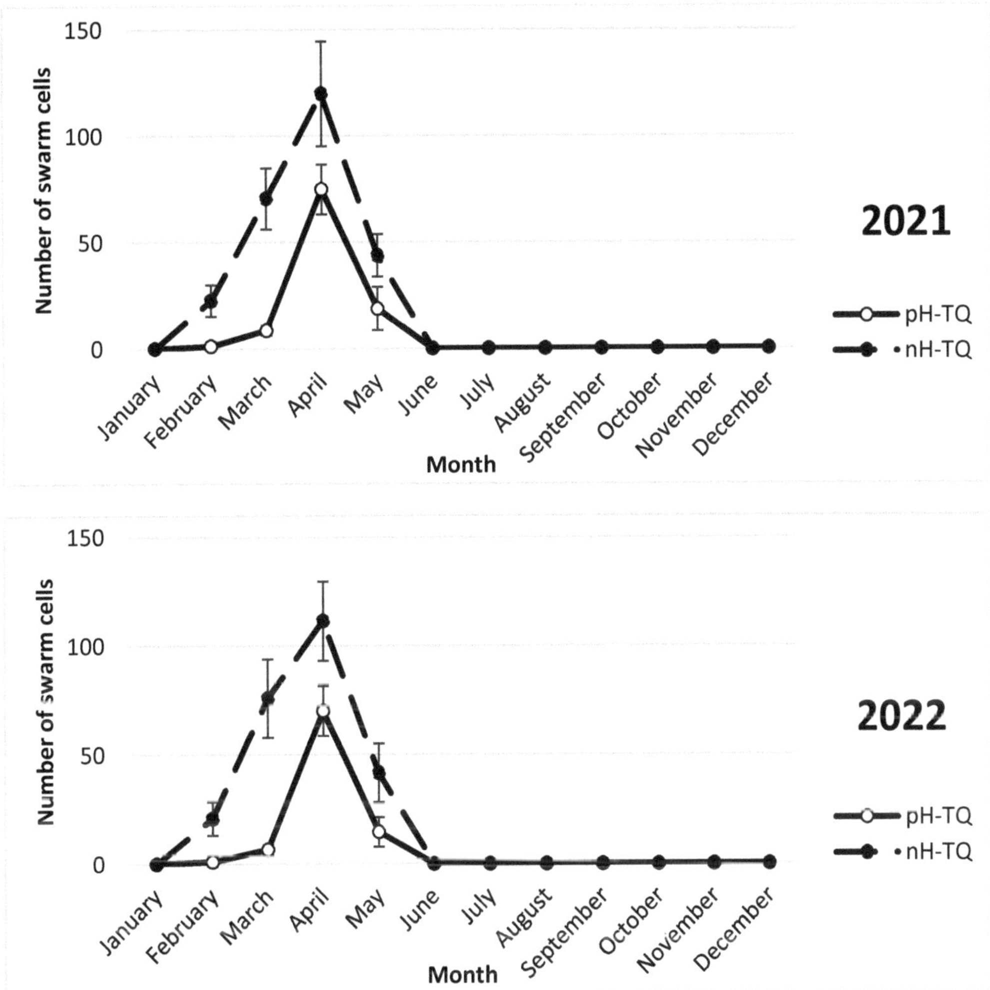

Figure 4. Number of swarm cell constructions (±standard deviation) of non-heat-treated honeybee queens (nH-TQ) and thermally manipulated honeybee queens in their immature stages (pH-TQ) (n = 12 colonies for both groups) in two consecutive years: 2021 and 2022.

3.5. Pollen Foraging Percentage per Hour

The workers in both colony groups, headed by the pH-TQs and nH-TQs, started foraging very early in the day at six o'clock (Figure 5). The pH-TQ colonies showed a significantly ($p < 0.001$) higher foraging rate from 8 AM to the end of the day than the nH-TQ colonies. A repeated-measures ANOVA showed a significant difference in the hours of flight time between the pH-MQ and nH-MQ groups ($F (1, 574) = 23.96$, $p < 0.001$).

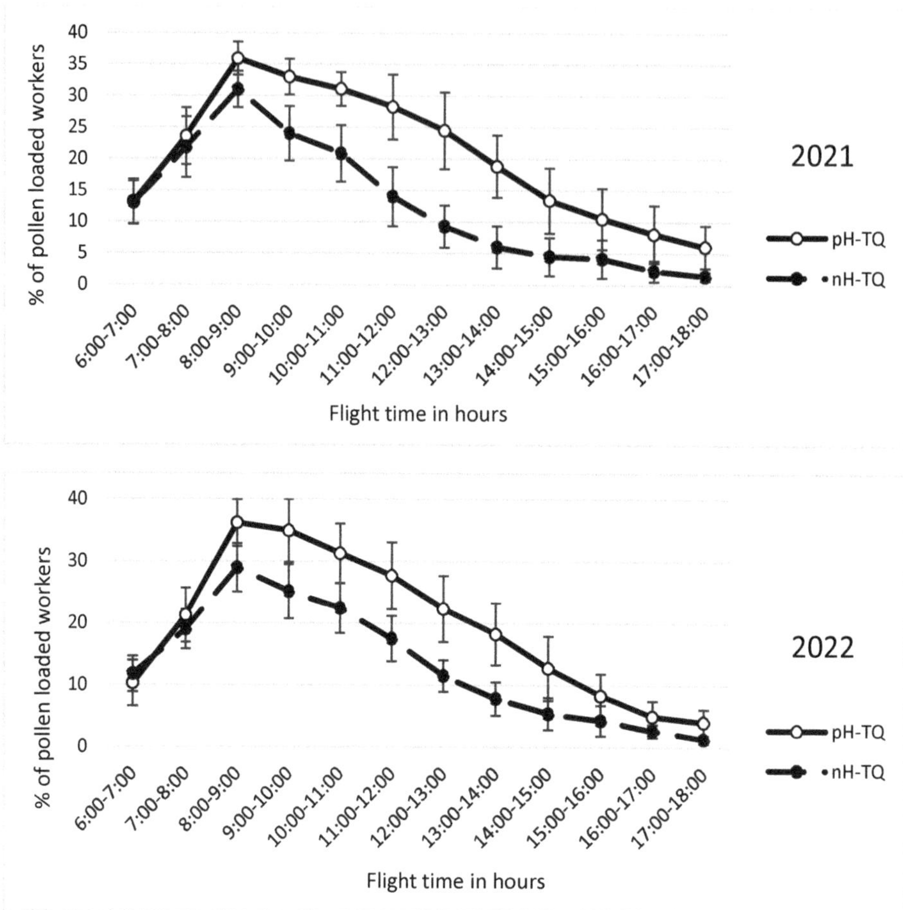

Figure 5. Percentage (%) of pollen loaded workers per hour (±standard deviation) of non-heat-treated honeybee queens (nH-TQ) and thermally manipulated honeybee queens in their immature stages (pH-TQ) in two consecutive years: 2021 and 2022.

3.6. Stored Pollen Cells

The number of stored pollen cells in both the pH-TQ and nH-TQ colonies fluctuated between the months of the year (Figure 6). The pH-TQ colonies stored significantly ($p < 0.001$) more pollen cells than the nH-TQ colonies in March (14,350 and 10,100 workers, respectively) and in May (12,167 and 8.275 cells, respectively). The pH-TQ colonies were more tolerant of high-temperature stress than the nH-TQ colonies, showing more active flying outside the hives and storing significantly more pollen cells ($p < 0.001$) during the summertime between May and August. In the autumn, when the air temperatures were more suitable and certain fruit trees began to flower, the pH-TQ colonies again stored significantly more pollen cells ($p < 0.001$) than the nH-TQ colonies (6238 and 1638 cells, respectively). In both colony groups, pollen storage was the lowest between December and February, when cold weather prevailed. A repeated-measures ANOVA showed a significant difference in the number of stored pollen cells between the pH-MQ and nH-MQ groups ($F(1, 574) = 64.93$, $p < 0.001$).

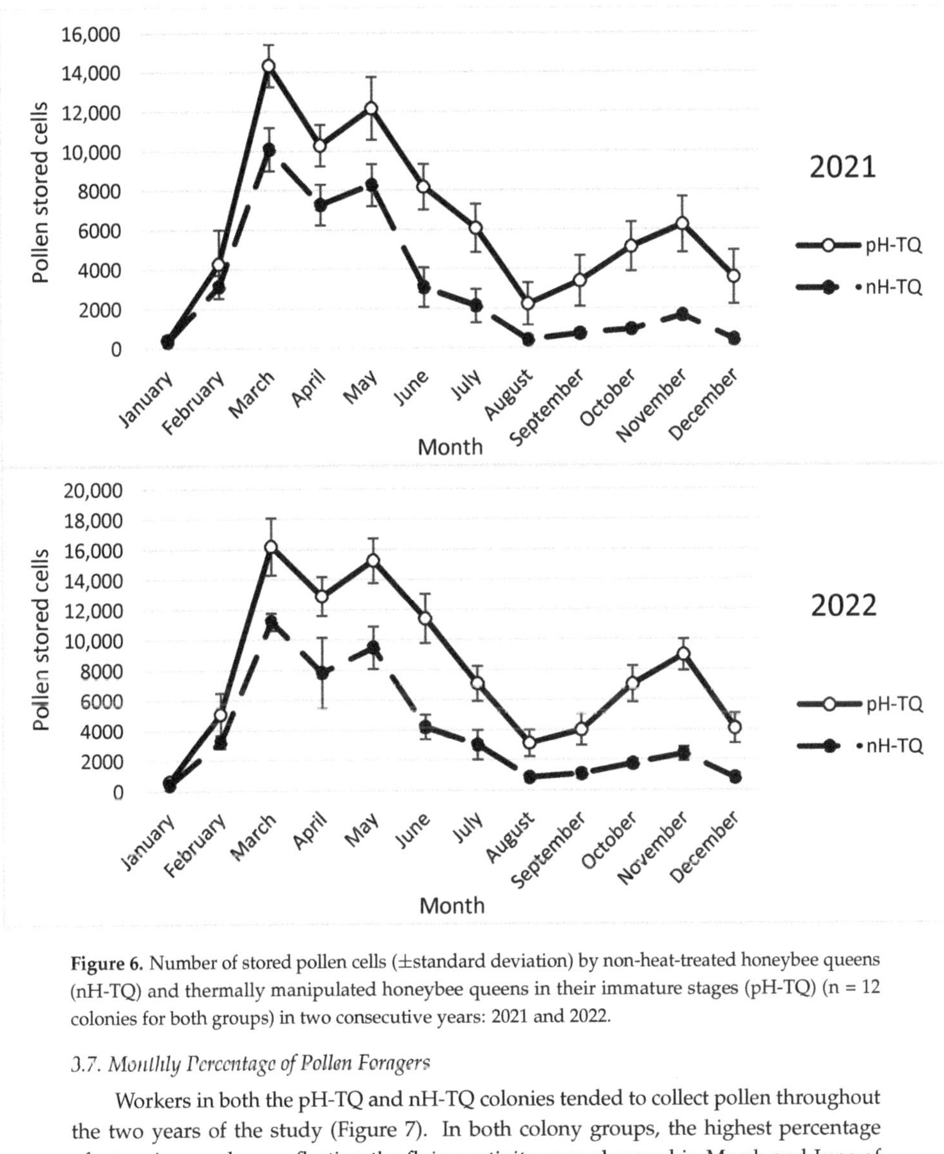

Figure 6. Number of stored pollen cells (±standard deviation) by non-heat-treated honeybee queens (nH-TQ) and thermally manipulated honeybee queens in their immature stages (pH-TQ) (n = 12 colonies for both groups) in two consecutive years: 2021 and 2022.

3.7. Monthly Percentage of Pollen Foragers

Workers in both the pH-TQ and nH-TQ colonies tended to collect pollen throughout the two years of the study (Figure 7). In both colony groups, the highest percentage of returning workers, reflecting the flying activity, was observed in March and June of each year. During the period from May to the end of the year, the average percentage of returning workers loaded with pollen grains in the pH-TQ colonies was significantly ($p < 0.001$) higher than that in the nH-TQ colonies. The forager percentages in the pH-TQ colonies were significantly ($p < 0.001$) higher in March, May, and October (76, 67.2, and 31.5%, respectively) than those in the nH-TQ colonies (61.2, 60.1, and 17.0%, respectively). A repeated-measures ANOVA showed a significant difference in the monthly pollen forager percentages between the pH-MQ and nH-MQ groups ($F(1, 574) = 31.88$, $p < 0.001$).

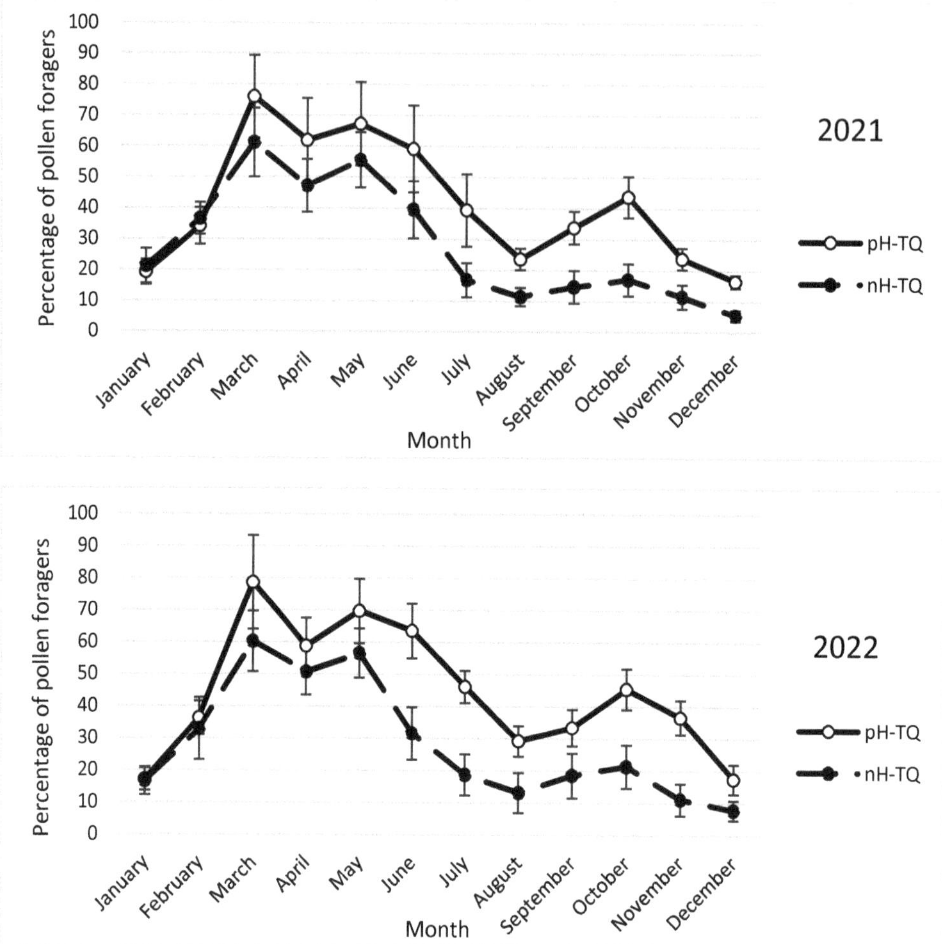

Figure 7. Percentage of returning workers collecting pollen (±standard deviation) in non-heat-treated honeybee queens (nH-TQ) and thermally manipulated honeybee queens in their immature stages (pH-TQ) in two consecutive years: 2021 and 2022.

3.8. Weight of 10 Pollen Loads

The average weight of 10 pollen loads carried by the workers in the pH-TQ colonies was significantly heavier than that carried by the workers in the nH-TQ colonies between the period from May to the end of the year (Figure 8). The heaviest pollen loads were recorded from March to May for both the pH-TQ and nH-TQ colonies. The workers in the pH-TQ colonies carried significantly ($p < 0.001$) heavier pollen loads in March, May, and October 2021 (195.1, 195, and 114.2 mg, respectively) than those in the nH-TQ colonies (165.1, 139.5, and 68.3 mg, respectively). A repeated-measures ANOVA showed a significant difference in the pollen load weight between the pH-MQ and nH-MQ groups ($F(1, 574) = 48.05$, $p < 0.001$).

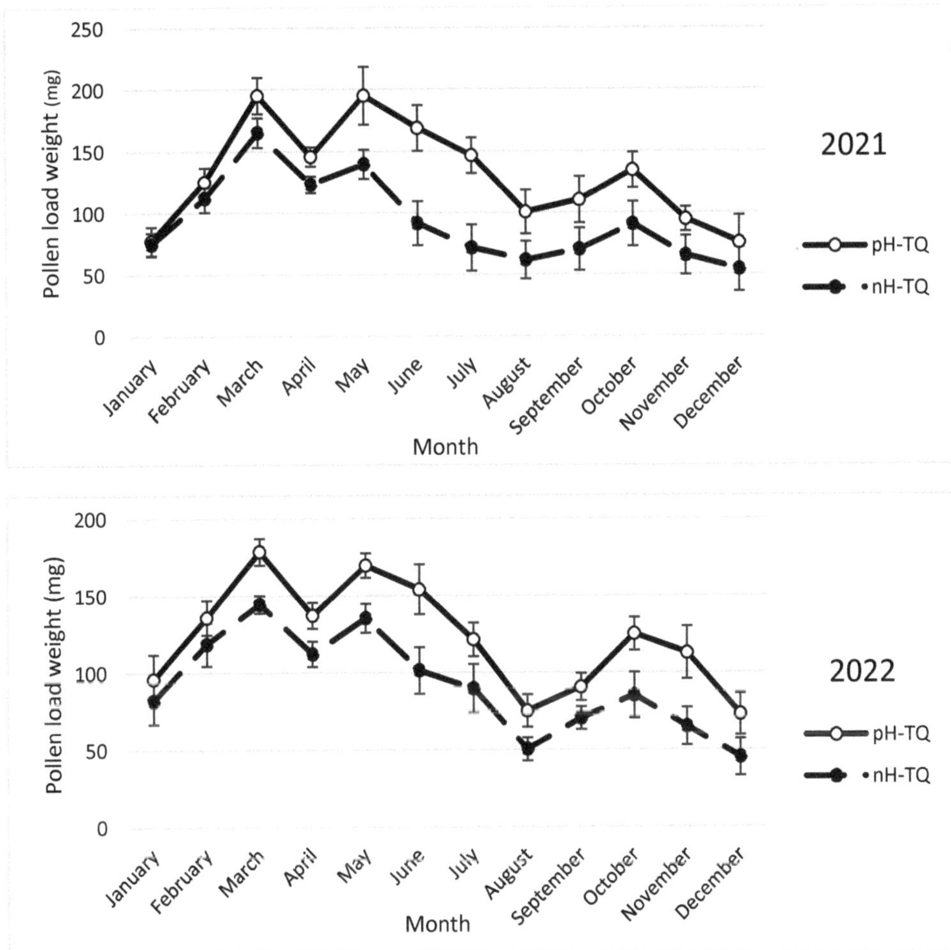

Figure 8. Average weight of 10 pollen loads in milligrams (mg) (±standard deviation) carried by workers of non-heat-treated honeybee queens (nH-TQ) and thermally manipulated honeybee queens in their immature stages (pH-TQ) in two consecutive years: 2021 and 2022.

In the second year of the study (2022), the trends in brood rearing, worker population, pollen collection and storage, swarm cell construction, and drone brood rearing in the pH-TQ and nH-TQ colonies tended to be the same and were consistent in terms of time (Figures 1–8). Supplementary Materials can be found in the link provided at the end of this article.

4. Discussion

4.1. Effect of Preheat Hardening on the Survival of Honeybee Workers

Overall, the results of the current investigation show that the preheat hardening of honeybee queens during one of their immature stages is a promising method for significantly improving the productive potential of honeybee colonies. Rapid heat hardening generally refers to a rapid improvement in survival at high lethal temperatures following brief pre-exposure to a sub-lethal temperature shock [10,11]. This study found that honeybee colonies headed by heat-treated queens (pH-TQs) and non-treated queens (nH-TQs) exhibited a triple cycle of brood rearing and adult populations within a year. The three peaks of the brood cells observed in March, May, and October of each year were comparable to those of the stored pollen inside the experimental colonies. The pollen foraging tendency of honeybee colonies depends on the genotype, seasonal availability, number of larvae, and amount of stored pollen [40,41]. The activity of pollen foragers depends on the flowering time of the plant species and the outside temperature [42]. A previous study found that an ambient temperature of 20 °C stimulated the highest foraging activity, while an ambient temperature of 43 °C resulted in the lowest activity [43]. Flying outside the hive under elevated temperatures may alter foraging behavior [44]. In addition to facing heat stress during their foraging activities, bees might also suffer from problems with their thorax, which is crucial for their ability to fly. Honeybee workers show prolonged flight only with thorax temperatures between 29 and 49 °C [45,46].

4.2. Effect of Preheat Hardening on Brood Numbers

For honeybees to successfully grow, reproduce, and build strong colonies, they must have access to enough flowering plant resources at the right time [47]. Honeybee foragers collect nectar, pollen, and water from flowering plants [48]. Importantly, pollen grains are the sole nitrogenous source of food for brood rearing, normal development, worker longevity, and sustaining the requirements of other members of the hive [49,50]. Here, the observed decline in brood rearing for a short period in April can be explained by the reduced foraging activities of the worker bees, and these can be attributed to the local high temperatures, the seasonal east sandstorm event, the low relative humidity from the end of March to April in the Jordan Valley, and fewer places for egg laying due to the storage of honey in the brood areas inside the hives. The second reduction in brood rearing was recorded starting from June, and it was presumably due to the abundant incoming nectar and pollen deposited in the brood area after transporting the bee colonies to the upland areas of Jordan. This activity may have restricted the number of cells available to the queen for ovipositing. A reasonable gradual decrease in incoming nectar and pollen was expected during the hot, summer period from July to September, with a simultaneous rapid decrease in brood-rearing activities. The third reduction in brood rearing was observed in November with a brood stop; this was probably because of the prevailing ambient cold conditions preventing the workers from flying effectively outside the hives, as well as the reduced flowering of wild and cultivated plants. These observations are in alignment with those previously reported under semiarid conditions in the same study area [51] and in other parts of the world [52].

4.3. Changes Induced by Preheat Hardening on Honeybee Workers

In this study, both experimental honeybee colonies headed by the pH-TQs and nH-TQs were subjected to the same environmental conditions of the mild spring months and hot dry weather conditions prevailing during the summers in Jordan. Similarly to other colonies, the workers in both groups developed mechanisms to regulate their thoracic temperature, prevent overheating, and remain active across wide ranges of air temperatures [53]. During the summer and later months from May to the end of the year, the colonies headed by the pH-TQs built significantly more brood areas, produced more workers, and collected more pollen than the colonies headed by the nH-TQs. These differences could be due to differences in innate mechanisms, such as changes in the expression of heat shock proteins (HSPs), an increased rate of heat loss, and a decreased rate of metabolic heat synthesis to prevent thorax overheating [53,54]. Moreover, the increased productivity of the pH-TQ colonies compared to that of the nH-TQ colonies may be attributed to the rapid synthesis of heat resistance substances, such as mannitol and sorbitol, both of which have been linked to increased heat tolerance in insects [55–57]. Any or all of these factors could explain the improvement in the natural foraging activities of the workers in the colonies headed by the preheated queens (pH-TQs), consequently increasing the amounts of nectar and pollen collected during the hot months and until the end of the year. This increased activity of pollen collection probably led to the increase in brood rearing and, subsequently, the increase in the worker population.

Previous studies have shown that rapid heat hardening in animals during the immature stages results in the production of individuals with improved body characteristics, represented by a larger body size, weight, and body organs [58–60]. These changes are further expressed in changes in the performance of the subsequent adult stages [61,62]. Examples of the effects of rapid heat hardening on the body morphology of insects are as follows: an increase in the body size of the ground cricket *Allonemobius socius*, the leafhopper *Scaphoideus titanus*, and the speckled wood butterfly (*Parage aegeria*) [62,63]; an increase in the body weight of *Harmonia axyridis* [64]; and an increase in the wing size of *Drosophila mercatorum* and the hoverfly *Eristalis arbustorum* [65,66]. The changes in the body morphology of worker honeybees due to heat hardening are currently under investigation by the authors.

The increase in the body size of insects may contribute to better heat tolerance and therefore increased productivity. In ants, an increase in the body size of workers can significantly improve their heat tolerance [67–69]. In bees, larger workers can reach flowering resources more quickly and over a greater distance than their smaller counterparts, thereby resulting in a better foraging scope and a higher pollen load [70,71]. Foraging for long distances with a better adaptability to hot weather conditions can explain the currently observed improved pollen collection of the pH-MQs. Furthermore, such an increased food supply inside the hive can accelerate the queens' egg-laying activity, thereby resulting in a larger worker population. Moreover, in accordance with metabolic scaling theory [72], the energy efficiency of insect flight increases with body mass [73–75], with larger insects exhibiting different flight kinematics expressed in different wingbeat frequencies [74,75]. Another important consideration here is the capacity of individuals to store food. Across insects, the absolute energy demand for flight is generally greater in larger-bodied individuals, which can be compensated for by greater baseline endogenous energy stores. This results in an increased flight endurance capacity with an increasing insect body size [76].

Resistance to extremely high ambient temperatures is achieved by a combination of qualities strongly affected by a prior-experienced environment [77]. Preheat hardening results in the enhanced expression of several heat shock proteins (HSPs) [56,78,79]. The expression of HSPs was found to be significantly associated with improved heat resistance [80,81] and might help prolong life [82], hence rendering the bees more tolerant to heat changes and better adapted to increased ambient temperatures. Additionally, preheat hardening was previously linked to the induction of immune gene expression [83], which might enhance the insects' responses to parasitic pathogens and stresses or unfavorable conditions [84]. Such an increase in the tolerance of bees might further lead to better flight foraging frequencies outside the hive. Additionally, previous studies showed that preheat hardening induces the expression of oxidative stress genes that diminish the adverse effects of free radicals, such as reactive oxygen species (ROS) [85,86]. The cellular scavenging of ROS was previously linked to a reduction in the oxidative stress imposed on honeybees, which was further linked to an increase in the lifespan of both honeybee workers and queens [87,88].

5. Conclusions

The current results show that exposing mother queens to non-lethal heat shock has positive effects on their reproduction and further improves the resilience of their worker daughters to environmental stressors. The increased ability of honeybee workers to collect nectar and pollen during the hot summer months indicates that they have a higher tolerance to ambient heat and allows them to overcome food shortages, which may compensate for the ongoing decline in bee populations locally and globally. In the context of beekeeping, we recommend that beekeepers expose honeybee queens to preheat hardening in their immature stages as a method to improve their subsequent daughter workers. This could be an effective way to maintain beehives and improve the heat tolerance of treated bees. Future studies examining the impact of preheat hardening on honeybee immunity and other factors involved in improving honeybee tolerance and adaptation to different stress conditions are recommended.

Supplementary Materials: The following supporting information can be downloaded at: https://www.mdpi.com/article/10.3390/insects16080858/s1, Table S1: Repeated Measures ANOVA Results for Hive Parameters (2021–2022). TM: thermally manipulated.

Author Contributions: Conceptualization, resources, visualization, supervision, funding acquisition, A.A.-M.A.-G. and S.T.Z.; methodology, validation, formal analysis, investigation, A.A.-M.A.-G., S.T.Z., I.M.A.-O., and M.S.A.; data curation, A.A.-M.A.-G., S.T.Z., M.N.S.A.-S., S.S.M., I.M.A.-O., and M.S.A.; writing—original draft preparation, writing—review and editing, A.A.-M.A.-G., S.T.Z., M.N.S.A.-S., and I.M.A.-O. All authors have read and agreed to the published version of the manuscript.

Funding: The authors thank the Deanship of Scientific Research at Jordan University of Science and Technology for funding the work through research grant no. 431/2021 and grant no. 386/2023.

Data Availability Statement: The original data presented in the study are openly available in https://drive.google.com/drive/folders/1o62SHEW9xL5OBC6ZxBEIT4BEi5fjOMkh?usp=sharing, (accessed on 22 May 2025).

Acknowledgments: We acknowledge Yousef Al-Ghzawi for performing the statistical analysis. We also acknowledge Al-Balqa Applied University, Salt, for their essential logistical support. We extend our acknowledgement to the reviewers and the editor whom their comments have improved the quality of this article.

Conflicts of Interest: The authors declare no conflicts of interest. The funders had no role in the design of the study; in the collection, analyses, or interpretation of data; in the writing of the manuscript; or in the decision to publish the results.

Abbreviation

The following abbreviations are used in this manuscript:

pH-TQ	Pre-heat-treated queens
nH-TQ	non-heat-treated queens
RH	Relative humidity

References

1. Vincent, C.; Hallman, G.; Panneton, B.; Fleurat-Lessard, F. Management of agricultural insects with physical control methods. *Annu. Rev. Entomol.* **2003**, *48*, 261–281. [CrossRef]
2. Beckett, S.; Fields, P.; Subramanyam, B. Disinfestation of stored products and associated structures using heat. In *Heat Treatments for Postharvest Pest Control: Theory and Practice*; CABI: Wallingford, UK, 2007; pp. 182–237.
3. Hansen, J.D.; Johnson, J.A.; Winter, D.A. History and use of heat in pest control: A review. *Int. J. Pest. Manag.* **2011**, *57*, 267–289. [CrossRef]
4. Grondeau, C.; Samson, R.; Sands, D.C. A Review of Thermotherapy to Free Plant Materials from Pathogens, Especially Seeds from Bacteria. *Crit. Rev. Plant Sci.* **1994**, *13*, 57–75. [CrossRef]

5. Wang, Y.; Li, S.; Zhang, Y.; Chen, Y.; Yan, F.; Han, L.; Ma, Y. Heat and cold therapy reduce pain in patients with delayed onset muscle soreness: A systematic review and meta-analysis of 32 randomized controlled trials. *Phys. Ther. Sport* **2021**, *48*, 177–187. [CrossRef] [PubMed]
6. Harbo, J.R. Heating adult honeybees to remove *Varroa jacobsoni*. *J. Apic. Res.* **2000**, *39*, 181–182. [CrossRef]
7. Goras, G.; Tananaki, C.; Gounari, S.; Dimou, M.; Lazaridou, E.; Karazafiris, E.; Kanelis, D.; Liolios, V.; El Taj, H.F.; Thrasyvoulou, A. Hyperthermia non-chemical control strategy against varroa. *J. Hell. Vet. Med. Soc.* **2015**, *66*, 249–256. [CrossRef]
8. Martin-Hernandez, R.; Meana, A.; Garcia-Palencia, P.; Marin, P.; Botias, C.; Garrido-Bailon, E.; Barrios, L.; Higes, M. Effect of temperature on the biotic potential of honeybee microsporidia. *Appl. Environ. Microbiol.* **2009**, *75*, 2554–2557. [CrossRef]
9. McMenamin, A.J.; Daughenbaugh, K.F.; Flenniken, M.L. The Heat Shock Response in the Western Honeybee (*Apis mellifera*) is Antiviral. *Viruses* **2020**, *12*, 245. [CrossRef]
10. Chidawanyika, F.; Terblanche, J.S. Rapid thermal responses and thermal tolerance in adult codling moth *Cydia pomonella* (Lepidoptera: Tortricidae). *J. Insect Physiol.* **2011**, *57*, 108–117. [CrossRef]
11. Mutamiswa, R.; Chidawanyika, F.; Nyamukondiwa, C. Superior basal and plastic thermal responses to environmental heterogeneity in invasive exotic stemborer Chilo partellus Swinhoe over indigenous *Busseola fusca* (Fuller) and *Sesamia calamistis* Hampson. *Physiol. Entomol.* **2018**, *43*, 108–119. [CrossRef]
12. Dalmon, A.; Peruzzi, M.; Le Conte, Y.; Alaux, C.; Pioz, M. Temperature-driven changes in viral loads in the honey bee *Apis mellifera*. *J. Invertebr. Pathol.* **2019**, *160*, 87–94. [CrossRef]
13. Li, X.; Ma, W.; Shen, J.; Long, D.; Feng, Y.; Su, W.; Xu, K.; Du, Y.; Jiang, Y. Tolerance and response of two honeybee species *Apis cerana* and *Apis mellifera* to high temperature and relative humidity. *PLoS ONE* **2019**, *14*, e0217921. [CrossRef]
14. Ahamed, M.; Posgai, R.; Gorey, T.J.; Nielsen, M.; Hussain, S.M.; Rowe, J.J. Silver nanoparticles induced heat shock protein 70, oxidative stress and apoptosis in *Drosophila melanogaster*. *Toxicol. Appl. Pharmacol.* **2010**, *242*, 263–269. [CrossRef]
15. Xu, H.; Miao, X.; Wang, W.; Wang, G.; Li, Y. Transcriptome analysis reveals the early resistance of zebrafish larvae to oxidative stress. *Fish Physiol. Biochem.* **2022**, *48*, 1075–1089. [CrossRef]
16. Elekonich, M.M. Extreme thermotolerance and behavioral induction of 70-kDa heat shock proteins and their encoding genes in honey bees. *Cell Stress Chaperones* **2009**, *14*, 219–226. [CrossRef] [PubMed]
17. Severson, D.W.; Erickson, E.H.; Williamson, J.L., Jr.; Aiken, J.M. Heat stress induced enhancement of heat shock protein gene activity in the honey bee (*Apis mellifera*). *Experientia* **1990**, *46*, 737–739. [CrossRef]
18. Ramirez, L.; Luna, F.; Mucci, C.A.; Lamattina, L. Fast weight recovery, metabolic rate adjustment and gene-expression regulation define responses of cold-stressed honey bee brood. *J. Insect Physiol.* **2021**, *128*, 104178. [CrossRef] [PubMed]
19. Degrandi-Hoffman, G.; Spivak, M.; Martin, J.H. Role of thermoregulation by nestmates on the development time of honey bee (Hymenoptera: Apidae) queens. *Ann. Entomol. Soc. Am.* **1993**, *86*, 165–172. [CrossRef]
20. Tautz, J.; Maier, S.; Groh, C.; Rössler, W.; Brockmann, A. Behavioral performance in adult honey bees is influenced by the temperature experienced during their pupal development. *Proc. Natl. Acad. Sci. USA* **2003**, *100*, 7343–7347. [CrossRef]
21. Abou-Shaara, H.; Owayss, A.A.; Ibrahim, Y.; Basuny, N. A review of the impacts of temperature and relative humidity on various activities of honey bees. *Insectes Sociaux* **2017**, *64*, 455–463. [CrossRef]
22. Chuda-Mickiewicz, B.; Samborski, J. The quality of honey bee queens from queen cells incubated at different temperatures. *Acta Sci. Pol. Zootech.* **2015**, *14*, 25–32.
23. Medina, R.G.; Paxton, R.J.; Hernandez-Sotomayor, S.M.T.; Pech-Jimenez, C.; Medina-Medina, L.A.; Quezada-Euan, J.J.G. Heat stress during development affects immunocompetence in workers, queens and drones of Africanized honey bees (*Apis mellifera* L.) (Hymenoptera: Apidae). *J. Therm. Biol.* **2020**, *89*, 102541. [CrossRef] [PubMed]
24. Langowska, A.; Zawilak, M.; Sparks, T.H.; Glazaczow, A.; Tomkins, P.W.; Tryjanowski, P. Long-term effect of temperature on honey yield and honeybee phenology. *Int. J. Biometeorol.* **2017**, *61*, 1125–1132. [CrossRef] [PubMed]
25. Jones, J.C.; Helliwell, P.; Beekman, M.; Maleszka, R.; Oldroyd, B.P. The effects of rearing temperature on developmental stability and learning and memory in the honey bee, *Apis mellifera*. *J. Comp. Physiol. A Neuroethol. Sens. Neural Behav. Physiol.* **2005**, *191*, 1121–1129. [CrossRef]
26. Alqarni, A.S. Tolerance of summer temperature in imported and indigenous honeybee *Apis mellifera* L. races in central Saudi Arabia. *Saudi J. Biol. Sci.* **2006**, *13*, 123–127. [CrossRef]
27. Blažytė-Čereškienė, L.; Vaitkevičienė, G.; Venskutonytė, S.; Būda, V. Honey bee foraging in spring oilseed rape crops under high ambient temperature conditions. *Zemdirbyste* **2010**, *79*, 61–70.
28. Le Conte, Y.; Navajas, M. Climate change: Impact on honey bee populations and diseases. *Revue Scientif. Techn.* **2008**, *27*, 499–510. [CrossRef]
29. Abou-Shaara, H.F.; Al-Ghamdi, A.A.; Mohamed, A.A. Tolerance of two honeybee races to various temperature and relative humidity gradients. *Environ. Exp. Biol.* **2012**, *10*, 133–138.
30. Stabentheiner, A.; Kovac, H.; Brodschneider, R. Honeybee colony thermoregulation–regulatory mechanisms and contribution of individuals in dependence on age, location and thermal stress. *PLoS ONE* **2010**, *5*, e8967. [CrossRef]

31. Southwick, E.E.; Moritz, R.F.A. Social control of air ventilation in colonies of honey bees, *Apis mellifera*. *J. Insect Physiol.* **1987**, *33*, 623–626. [CrossRef]
32. Nicolson, S.W. Water homeostasis in bees, with the emphasis on sociality. *J. Exp. Biol.* **2009**, *212*, 429–434. [CrossRef]
33. Stabentheiner, A.; Kovac, H.; Mandl, M.; Käfer, H. Coping with the cold and fighting the heat: Thermal homeostasis of a superorganism, the honeybee colony. *J. Comp. Physiol.* **2021**, *207*, 337–351. [CrossRef]
34. Starks, P.T.; Gilley, D.C. Heat shielding: A novel method of colonial thermoregulation in honey bees. *Naturwissenschaften* **1999**, *86*, 438–440. [CrossRef]
35. Al-Ghzawi, A.A.A.; Zaitoun, S. Origin and Rearing Season of Honeybee Queens Affect Some of Their Physiological and Reproductive Characteristics. *Entomol. Res.* **2008**, *38*, 139–148. [CrossRef]
36. Laidlaw, H.H.; Page, R.E., Jr. *Queen Rearing and Bee Breeding*, 1st ed.; Wicwas Press: Cheshire, CT, USA, 1997; p. 224.
37. Gerig, L. Lehrgang zur erfassung der volksstarke. *Schweiz. Bienenztg.* **1983**, *106*, 199–204.
38. Ruttner, H. Technical recommendations for methods of evaluating performance of bee colonies. In *Controlled Mating and Selection of the Honey Bee*; Ruttner, F., Ed.; Apimondia: Bucharest, Romania, 1972; pp. 87–92.
39. R Core Team. *R: A Language and Environment for Statistical Computing (Computer Software)*; R Foundation for Statistical Computing: Vienna, Austria, 2021.
40. Camazine, S. The regulation of pollen foraging by honey bees: How foragers assess the colony's need for pollen. *Behav. Ecol. Sociobiol.* **1993**, *32*, 265–272. [CrossRef]
41. Bilisik, A.; Cakmak, I.; Bicakci, A.; Malyer, H. Seasonal variation of collected pollen loads of honeybees (*Apis mellifera* L. anatoliaca). *Grana* **2008**, *47*, 70–77. [CrossRef]
42. Ghosh, S.; Jeon, H.; Jung, C. Foraging behaviour and preference of pollen sources by honey bee (*Apis mellifera*) relative to protein contents. *J. Ecol. Environ.* **2020**, *44*, 4. [CrossRef]
43. Tan, K.; Yang, S.; Wang, Z.-W.; Radloff, S.E.; Oldroyd, B.P. Differences in foraging and broodnest temperature in the honey bees *Apis cerana* and *A. mellifera*. *Apidologie* **2012**, *43*, 618–623. [CrossRef]
44. Cooper, P.D.; Schaffer, W.M.; Buchmann, S.L. Temperature Regulation of Honey Bees (*Apis mellifera*) Foraging in the Sonoran Desert. *J. Exp. Biol.* **1985**, *114*, 1–15. [CrossRef]
45. Heinrich, B. Resource heterogeneity and patterns of movement in foraging bumblebees. *Oecologia* **1979**, *40*, 235–245. [CrossRef] [PubMed]
46. Coelho, J.R. Heat Transfer and Body Temperature in Honey Bee (Hymenoptera: Apidae) Drones and Workers. *Environ. Entomol.* **1991**, *20*, 1627–1635. [CrossRef]
47. Brodschneider, R.; Crailsheim, K. Nutrition and health in honey bees. *Apidologie* **2010**, *41*, 278–294. [CrossRef]
48. Thorp, R.W. The collection of pollen by bees. *Plant Syst. Evol.* **2000**, *222*, 211–223. [CrossRef]
49. Di Pasquale, G.; Alaux, C.; Le Conte, Y.; Odoux, J.F.; Pioz, M.; Vaissiere, B.E.; Belzunces, L.P.; Decourtye, A. Variations in the Availability of Pollen Resources Affect Honey Bee Health. *PLoS ONE* **2016**, *11*, e0162818. [CrossRef]
50. Mattila, H.R.; Otis, G.W. Influence of pollen diet in spring on development of honey bee (Hymenoptera: Apidae) colonies. *J. Econ. Entomol.* **2006**, *99*, 604–613. [CrossRef]
51. Zaitoun, S.T.; Al-Ghzawi, A.-M.; Shannag, H.K. Population dynamics of the Syrian Honeybee, *Apis mellifera* syriaca, under semi-arid Mediterranean conditions. *Zool. Middle East* **2000**, *21*, 129–132. [CrossRef]
52. Jeffree, E.P. Observations on the Decline and Growth of Honey Bee Colonies1. *J. Econ. Entomol.* **1955**, *48*, 723–726. [CrossRef]
53. Roberts, S.P.; Harrison, J.F. Mechanisms of thermal stability during flight in the honeybee *Apis mellifera*. *J. Exp. Biol.* **1999**, *202*, 1523–1533. [CrossRef]
54. Perez, R.; Aron, S. Adaptations to thermal stress in social insects: Recent advances and future directions. *Biol. Rev. Camb. Philos. Soc.* **2020**, *95*, 1535–1553. [CrossRef]
55. Colinet, H.; Lee, S.F.; Hoffmann, A. Temporal expression of heat shock genes during cold stress and recovery from chill coma in adult *Drosoph melanogaster*. *FEBS J.* **2010**, *277*, 174–185. [CrossRef]
56. Al-Ghzawi, A.A.A.; Al-Zghoul, M.B.; Zaitoun, S.; Al-Omary, I.M.; Alahmad, N.A. Dynamics of Heat Shock Proteins and Heat Shock Factor Expression During Heat Stress in Daughter Workers in Pre-Heat-Treated (Rapid Heat Hardening) *Apis mellifera* Mother Queens. *J. Therm. Biol.* **2022**, *104*, 103194. [CrossRef]
57. Zaitoun, S.T.; Al-Ghzawi, A.A.; Al-Zghoul, M.B.; Al-Omary, I.M.; Al-Sabi, M.N.S. Effects of Rapid Heat Hardening of Honeybee Queens (*Apis mellifera*) During the Larval Stage on the Oxidative Response of Their Workers During Heat Stress. *Stresses* **2025**, *5*, 32. [CrossRef]
58. Yahav, S.; Plavnik, I. Effect of early-stage thermal conditioning and food restriction on performance and thermotolerance of male broiler chickens. *Br. Poult. Sci.* **1999**, *40*, 120–126. [CrossRef]
59. Yahav, S.; McMurtry, J.P. Thermotolerance acquisition in broiler chickens by temperature conditioning early in life--the effect of timing and ambient temperature. *Poult. Sci.* **2001**, *80*, 1662–1666. [CrossRef]

60. Jonsson, B.; Jonsson, N.; Finstad, A.G. Linking embryonic temperature with adult reproductive investment in Atlantic salmon *Salmo salar*. *Mar. Ecol. Prog. Ser.* **2014**, *515*, 217–226. [CrossRef]
61. Gray, E.M. Thermal acclimation in a complex life cycle: The effects of larval and adult thermal conditions on metabolic rate and heat resistance in *Culex pipiens* (Diptera: Culicidae). *J. Insect Physiol.* **2013**, *59*, 1001–1007. [CrossRef]
62. Sgro, C.M.; Terblanche, J.S.; Hoffmann, A.A. What Can Plasticity Contribute to Insect Responses to Climate Change? *Annu. Rev. Entomol.* **2016**, *61*, 433–451. [CrossRef]
63. Stevens, D.J. Pupal development temperature alters adult phenotype in the speckled wood butterfly, *Pararge aegeria*. *J. Therm. Biol.* **2004**, *29*, 205–210. [CrossRef]
64. Chuche, J.; Thiery, D. Egg incubation temperature differently affects female and male hatching dynamics and larval fitness in a leafhopper. *Ecol Evol.* **2012**, *2*, 732–739. [CrossRef] [PubMed]
65. Zhang, S.; Cao, Z.; Wang, Q.; Zhang, F.; Liu, T.-X. Exposing eggs to high temperatures affects the development, survival and reproduction of *Harmonia axyridis*. *J. Therm. Biol.* **2014**, *39*, 40–44. [CrossRef]
66. Ottenheim, M.M.; Volmer, A.D. Wing length plasticity in *Eristalis arbustorum* (Diptera: Syrphidae). *Neth. J. Zool.* **1999**, *49*, 15–27. [CrossRef]
67. Andersen, D.H.; Pertoldi, C.; Scali, V.; Loeschcke, V. Heat stress and age induced maternal effects on wing size and shape in parthenogenetic *Drosophila mercatorum*. *J. Evol. Biol.* **2005**, *18*, 884–892. [CrossRef]
68. Baudier, K.M.; Mudd, A.E.; Erickson, S.C.; O'Donnell, S. Microhabitat and body size effects on heat tolerance: Implications for responses to climate change (army ants: Formicidae, Ecitoninae). *J. Anim. Ecol.* **2015**, *84*, 1322–1330. [CrossRef]
69. Wendt, C.F.; Verble-Pearson, R. Critical thermal maxima and body size positively correlate in red imported fire ants, *Solenopsis invicta*. *Southwest. Nat.* **2016**, *61*, 79–83. [CrossRef]
70. Greenleaf, S.S.; Williams, N.M.; Winfree, R.; Kremen, C. Bee foraging ranges and their relationship to body size. *Oecologia* **2007**, *153*, 589–596. [CrossRef]
71. Willmer, P.; Finlayson, K. Big bees do a better job: Intraspecific size variation influences pollination effectiveness. *J. Pollinat. Ecol.* **2014**, *14*, 244–254. [CrossRef]
72. Brown, J.H.; Gillooly, J.F.; Allen, A.P.; Savage, V.M.; West, G.B. Toward a metabolic theory of ecology. *Ecology* **2004**, *85*, 1771–1789. [CrossRef]
73. Skandalis, D.A.; Darveau, C.A. Morphological and physiological idiosyncrasies lead to interindividual variation in flight metabolic rate in worker bumblebees (*Bombus impatiens*). *Physiol. Biochem. Zool.* **2012**, *85*, 657–670. [CrossRef]
74. Billardon, F.; Darveau, C.A. Flight energetics, caste dimorphism and scaling properties in the bumblebee, *Bombus impatiens*. *J. Exp. Biol.* **2019**, *222*, jeb187807. [CrossRef]
75. Darveau, C.A.; Hochachka, P.W.; Welch, K.C.; Roubik, D.W., Jr.; Suarez, R.K. Allometric scaling of flight energetics in Panamanian orchid bees: A comparative phylogenetic approach. *J. Exp. Biol.* **2005**, *208*, 3581–3591. [CrossRef] [PubMed]
76. Kaufmann, C.; Reim, C.; Blanckenhorn, W.U. Size-dependent insect flight energetics at different sugar supplies. *Biol. J. Linn. Soc.* **2013**, *108*, 565–578. [CrossRef]
77. Wos, G.; Willi, Y. Thermal Acclimation in *Arabidopsis lyrata*: Genotypic Costs and Transcriptional Changes. *J. Evol. Biol.* **2018**, *31*, 123–135. [CrossRef] [PubMed]
78. Chen, P.C.; Lee, R.E.; Denlinger, D.L. Cold shock and heat shock: A comparison of the protection generated by brief pretreatment at less severe temperatures. *Physiol. Entomol.* **1991**, *16*, 19–26. [CrossRef]
79. Alqarni, A.S.; Ali, H.; Iqbal, J.; Owayss, A.A.; Smith, B.H. Expression of heat shock proteins in adult honey bee (*Apis mellifera* L.) workers under hot-arid subtropical ecosystems. *Saudi J. Biol. Sci.* **2019**, *26*, 1372–1376. [CrossRef]
80. Malmendal, A.; Overgaard, J.; Bundy, J.G.; Sorensen, J.G.; Nielsen, N.C.; Loeschcke, V.; Holmstrup, M. Metabolomic Profiling of Heat Stress: Hardening and Recovery of Homeostasis in *Drosophila*. *Am. J. Physiol. Regul. Integr. Comp. Physiol.* **2006**, *291*, R205–R212. [CrossRef]
81. Robert, J. Evolution of heat shock protein and immunity. *Dev. Comp. Immunol.* **2003**, *27*, 449–464. [CrossRef]
82. Feder, M.E.; Hofmann, G.E. Heat-shock proteins, molecular chaperones, and the stress response: Evolutionary and ecological physiology. *Annu. Rev. Physiol.* **1999**, *61*, 243–282. [CrossRef] [PubMed]
83. Blasco-Lavilla, N.; García-Reina, A.; De la Rúa, P. Mild thermal stress does not negatively affect immune gene expression in the bumblebee *Bombus terrestris*. *Apidologie* **2021**, *52*, 163–173. [CrossRef]
84. Larsen, A.; Reynaldi, F.J.; Guzmán-Novoa, E. Fundaments of the honey bee (*Apis mellifera*) immune system. Review. *Rev. Mex. Cienc. Pecu.* **2019**, *10*, 705–728. [CrossRef]
85. Corona, M.; Robinson, G.E. Genes of the Antioxidant System of the Honey Bee: Annotation and Phylogeny. *Insect Mol. Biol.* **2006**, *15*, 687–701. [CrossRef] [PubMed]
86. Santos, D.B.; Colle, D.; Moreira, E.L.G.; Santos, A.A.; Hort, M.A.; Santos, K.; Oses, J.P.; Razzera, G.; Farina, M. Probucol Protects Neuronal Cells Against Peroxide-Induced Damage and Directly Activates Glutathione Peroxidase-1. *Mol. Neurobiol.* **2020**, *57*, 3245–3257. [CrossRef]

87. Aurori, C.M.; Buttstedt, A.; Dezmirean, D.S.; Marghitas, L.A.; Moritz, R.F.; Erler, S. What Is the Main Driver of Ageing in Long-Lived Winter Honeybees: Antioxidant Enzymes, Innate Immunity, or Vitellogenin? *J. Gerontol. A Biol. Sci. Med. Sci.* **2014**, *69*, 633–639. [CrossRef] [PubMed]
88. Paleolog, J.; Wilde, J.; Miszczak, A.; Gancarz, M.; Strachecka, A. Antioxidation Defenses of *Apis mellifera* Queens and Workers Respond to Imidacloprid in Different Age-Dependent Ways: Old Queens Are Resistant, Foragers Are Not. *Animals* **2021**, *11*, 1246. [CrossRef] [PubMed]

insects

MDPI

Article

Living Control Systems: Exploring a Teleonomic Account of Behavior in *Apis mellifera*

Ian T. Jones [1], James W. Grice [2,*] and Charles I. Abramson [2]

[1] Applied Research and Analysis Company, Richmond, VA 23233, USA; ijones@aracscience.com
[2] Department of Psychology, Oklahoma State University, Stillwater, OK 74078, USA; charles.abramson@okstate.edu
* Correspondence: james.grice@okstate.edu

Simple Summary: We investigated how honey bees (*Apis mellifera*) adapt their behavior when navigating to a sugar-rich food source in the presence of wind that disrupts their flight path. To explain this behavior, we applied Perceptual Control Theory (PCT), which posits that organisms actively pursue goals rather than merely respond to external stimuli. In our experiment, 13 of 14 bees successfully adjusted their flight paths to reach the food source despite the wind. The bees tested various approach strategies but ultimately favored flying directly into the wind, rather than allowing it to push them from behind or the side. Consistent with PCT principles, our findings indicate that honey bee behavior is self-regulating and adaptable, resembling a dynamic adjustment process rather than a linear cause-and-effect sequence. As PCT emphasizes the purpose of behavior, it can also be described as goal-directed (teleonomic).

Abstract: Self-regulatory foraging behavior in honey bees (*Apis mellifera*) was investigated using the framework of Perceptual Control Theory (PCT). We developed a PCT-based model to describe how bees maintain goal-directed behavior, specifically targeting a sucrose-rich feeding site while overcoming a wind disturbance. In a controlled experiment, we found that 13 of 14 bees could successfully adjust their flight paths to overcome the disturbance and consistently reach the feeding target. While they demonstrated a great deal of individual variability regarding how they overcame the wind across experimental trials, they did so by finally adopting a headwind (i.e., flying into the wind) approach pattern rather than tailwind or crosswind approach patterns. These results support the application of PCT to the study of behavior in honey bees, which can be regarded as self-regulative (i.e., non-linear and dynamic) rather than as linear sequences of inputs and outputs. Given that such dynamic models are concerned with the functions or purposes of behavior, they may also be classified as teleonomic.

Keywords: self-regulation; teleonomy; behavior; honey bees; insects; organism-centered; person-centered

1. Introduction

1.1. Living Control Systems: Exploring a Teleonomic Account of Behavior in Apis mellifera

Self-regulatory systems play critical roles in the physiological processes of organisms. These systems work to maintain specific homeostatic set-points, such as body temperature or blood glucose levels, which are essential for an organism's optimal functioning and survival [1]. Regulatory systems also play important roles in governing the behaviors of organisms as they dynamically modify their actions on the basis of internal states and environmental stimuli [2]. Consider a common housefly, for example, perched on a pedestal in the middle of a drum marked with evenly spaced vertical lines. When the drum is rotated at different speeds by the experimenter, the fly will continuously adjust its behavior by rotating its body in the same direction and at the same speed as the drum, essentially maintaining an internal set-point of zero discrepancy between its movement and that of the drum [3]. As another example, consider a rat pressing a lever to receive a food pellet as a reward. As the experimenter increases the number of presses required to earn the reward, the rat will increase its rate of lever pressing accordingly while still maintaining a set rate at which it receives the reward [3]. In both examples self-regulatory systems are at work because the observed behaviors of the fly and rat are the result of a dynamic interplay between internal states and changing environmental stimuli. In the current paper we show how such systems can be modeled using the tools of Perceptual Control Theory and empirically studied in honey bees (*Apis mellifera*). We will also argue that self-regulatory behavioral systems are teleonomic and cannot be accurately understood otherwise.

1.2. Perceptual Control Theory

Explaining self-regulatory behavior is a particularly challenging task because it requires consideration of both the internal state of the organism as well as the feedback it receives from the environment. Simple behaviorist models run the risk of ignoring the former, while simple cognitive models run the risk of ignoring the latter. By "simple models" we here mean linear causation, input–output models (e.g., $S \to O \to R$) that fail to capture the complex and dynamic nature of regulatory systems. Examples and critiques of such models can be found in [2,4–6]. In more sophisticated language simple models can be said to ignore the function of behavior, which is to say they ignore the teleonomic aspects of behavior. William (Bill) Powers sought to correct the shortcomings of these input–output models with the introduction of his Perceptual Control Theory in the 1960s [5,7–9]. Building on cybernetic theory (e.g., [10–13]), Powers developed an intricate approach to modeling behavior with negative feedback loops which could also be organized into hierarchies to explain complex behavior. He explicitly made room in his approach for both the environment and the internal state of an organism, understanding behavior as an organism controlling its perceptions while it interacts with and receives feedback from the environment. Powers also regarded behavior as fundamentally teleonomic: "Behavior can be seen as purposive or goal-directed if it is recognized that the purpose of any act is to resist disturbances and that the reference condition describes the goal of the behavior." ([5], p. 54).

The essential features of a simple negative feedback loop can best be demonstrated via an artifact, namely, a thermostat. The input device on the modern thermostat senses and analyzes the current temperature in the environment and computes whether the current temperature of the environment matches the 'internalized' temperature set by the homeowner. In moments when the room temperature does not match the set temperature, an error signal is computed, and an electrical signal is sent by the thermostat to the heating/cooling unit which warms or cools the air. The thermostat continues monitoring the air and compares the room and set temperature values, turning off the heating/cooling unit when the two values match. From the perspective of Perceptual Control Theory, the thermostat is a negative feedback control system, the function—or teleonomy—of which is to regulate the temperature in the room.

Figure 1 shows a model developed by Powers [5] as a generic and adaptable means for visualizing such control systems. Applying this model to the thermostat, the "Input Function" is the sensor on the thermostat that generates a value (p) sent to the "Comparator." The "Reference Signal" (r) in the model is the temperature set by the homeowner, and the "Error Signal" (e) is the discrepancy between the room's current temperature and the set value as determined by the comparator. The "Output Function" transforms the error signal which is sent to the heating/cooling unit as the "Output Quantity" (o) which determines its action (heat, cool, off). The action of the heating/cooling unit impacts the environment, the "Environmental Function," and changes the temperature, the "Input Quantity" (i). The input quantity, then, is the moment-to-moment room temperature, and the "Disturbance" (d) is anything which can act upon the input quantity (e.g., outside weather, a hot oven, an open window). The negative feedback relationship continues so long as error is produced between the perceptual signal and referent signal. So long as the room temperature does not match the set temperature on the thermostat, the heating/cooling unit will act on the environment and counteract the disturbances.

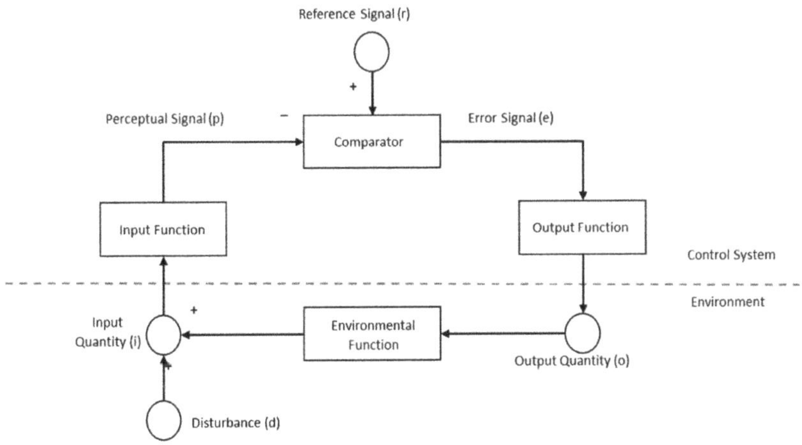

Figure 1. A Generic negative feedback loop for Perceptual Control Theory.

What the control model shows is that the observed output to the environment (i.e., the cold or hot air blowing) is not a result of the input (i.e., elevated room temp) alone, but is rather the result of the discrepancy between the input signal (i.e., the current temperature) and the desired temperature (i.e., the pre-specified temperature); in other words, there is a negative feedback relationship between the inputs and the outputs. A linear causal model is simply not sufficient for explaining the behavior of the thermostat nor the observed changes in the variables measured in the system. Powers [5] believed that such negative feedback relationships can similarly be used to explain the simple and complex behaviors of organisms. Unlike machines, however, which have values programmed into them by external agents (e.g., a homeowner setting a thermostat), the organisms themselves maintain the prespecified values and defend those values from disturbances (i.e., things in the environment) which would otherwise cause a misalignment between the actual and ideal values. A given behavior is therefore not to be understood as a linear chain of inputs and outputs, but rather as an output produced by the organism to defend or counteract disturbances which produce misalignments (i.e., errors) to internalized variables (i.e., the controlled variables).

Even relatively simple behaviors such as the common knee jerk can be understood from the perspective of Perceptual Control Theory. As described by Marken [2], a teleonomic explanation of this behavior is that the quadricep muscle has a predetermined state of relaxation or stretch when at rest at a 90° angle. When the doctor strikes the tendon while the leg is in this position, the quadricep muscle's state of stretch is disturbed. In other words, the hammer strike disturbs the degree of stretch in the quadricep muscle, and as a result the muscle resists it by contracting the quadricep and relaxing the hamstring. The resultant leg kick occurs because the muscle is resisting disturbance, or the stretching of the muscle produced by the tap (see [2,14], for more details). What appears to be a simple stimulus-response sequence of events is instead a control system at work.

In Perceptual Control Theory, the focus rests upon identifying the controlled variables, which are the variables the organism seeks to preserve at specified levels or amounts. The controlled variables have established reference signals, which are the pre-specified levels (i.e., set-points) at which they should be maintained. The controlled variables are identified when the organism behaves to counteract a disturbance, meaning that a particular controlled variable can only be identified by disturbing it. If the organism changes its behavior due to a disturbance, then the variable identified can be said to be 'under control' and labeled as a controlled variable.

Perceptual Control Theory has been successfully used to explain behaviors across a wide variety of species, including humans [9,15,16], rats [2,3,17–20], flies [3], German yellow jackets [21], crickets [18], jelly fish [22], and flat worms [23] as well as in contests between animals from the same species [24].

1.3. Current Experiment

In this study, we modified an operant conditioning paradigm to demonstrate how Perceptual Control Theory can be used to explain aspects of the foraging behavior of honey bees. Specifically, we trained bees to land on a rewarding target and then introduced a disturbance (viz., wind) to impede their approach to the target. The rewarding target was a 50% (by volume) sucrose solution placed on an orange dot. As a species, honey bees are foragers, meaning they must search for their food if they are to survive and reproduce. In accord with their nature, we trained the bees to repeatedly fly to, land upon, and feed from a rich food source.

From the perspective of Perceptual Control Theory, the distance between a bee and the food source is a control variable in which the ideal value is zero. In Figure 1 this is the Reference Signal (internal standard) of the bee. As the bee approaches its target, the Comparator returns a relatively large Error Signal, and various physiological mechanisms are consequently activated (Output Function) to produce observable flight (measured velocity, altitude, and direction; Output Quantities) toward the target (Environmental Function) as the measured distance (Input Quantity) is reduced. Our introduction of wind is a Disturbance the bee must overcome to reach the target. As shown in Figure 1, the Disturbance is part of the environment that affects the Input Function, the Perceptual Signal, and then the Comparator (completing the loop). To test for the presence of the control variable, we systematically altered the direction of the wind and recorded the behaviors of the honey bees. In accord with Perceptual Control Theory, we expected the bees to vary individually in the ways they overcame the disturbance, but to behave predictably and consistently in ultimately reaching the rewarding feeding target (i.e., the goal).

2. Materials and Methods

2.1. Subjects

The subjects were 14 experimentally naive honey bees (*Apis mellifera ligustica*) from two maintained hives located in central Oklahoma. The two colonies were housed in a standard two-box Langstroth hive consisting of approximately 20,000–40,000 bees. The hives were located in a pesticide-free area, and no chemicals were applied to the colonies for mite control. Bees from these two colonies and perhaps other bees in the vicinity regularly foraged at the feeder which was placed 44 feet (13.41 m) from the colonies. The colonies were approximately two years old and were obtained from an apiary. During the experiment, one of the colonies absconded. However, we are under the impression that bees from the absconded colony were still foraging on the feeder, as the number of foraging bees did not appear to decrease. Individual foragers were grabbed from a feeding station (see Figure 2) in a small matchbox and brought to a separate table. Throughout this experiment the bees were treated in accord with the American Zoological Association's ethical guidelines for research with animals. Oklahoma State University does not require an Institutional Animal Care and Use Committee (IACUC) review for research conducted with invertebrate animals.

Figure 2. Free flying honey bee feeding station. Note. The feeding station was 13.41 m away from the hive and 5.49 m away from the training/experiment table. The mason jars contain an 8% (by volume) sucrose solution.

2.2. Apparatus

The landing site for the experiment was placed atop a white circular table approximately 18.29 m from the hive and 5.49 m from the feeding station. As can be seen in Figure 3, the landing site was a 45 cm × 31.5 cm × 1.8 cm wooden board placed on top of an upside-down, white plastic basket (dimensions: 25.8 cm × 16.3 cm × 5.9 cm). The wooden cutting board was covered in white contact paper (brand: EasyLiner Contact Paper) to maintain a neutrally colored landing site. The bees fed from a 6 cm (diameter) clear, circular Petri dish. For the initial shaping trials a 6 cm (diameter) laminated gray (brand: Glidden, Color and Code: Granite Grey, 00NN 37/000Glidden, Pittsburgh, PA, USA) target was affixed to the bottom of the Petri dish, and a 25 mL droplet of 50% sucrose solution was placed on the center of the gray target. For the three control trials and nine experimental trials the gray target was replaced with a 4.5 cm (in diameter) circular, orange (brand: Valspar color chip, Autumn Blaze, #2002-1A; hex: #e55b2b, Valspar Corporation, Cleveland, OH, USA) target upon which the sucrose solution was placed.

Figure 3. Experimental testing table setup. Note. The training/experiment table was a white circular table that is approximately 18.29 m away from the hive and 5.49 m away from the feeding station (see Figure 2). The placement of the fan is always in reference to the hive (featured). The orange target is shown in the Petri dish.

The landing site was positioned in reference to the hive, such that the back of the landing site was always facing the hive (see Figure 3). The target was placed 12.6 cm away from the left and right sides, and 19.5 cm away from the front (as measured from the center of the target to the edge of the landing site). In addition, one small, black, circular, stroller fan (brand: AMACOOL, Shenzhen LuanSheng Electronic Technology Co., Ltd, Shenzhen, China) was placed 20 cm away from the center of the target in one of three different positions (see below). The fan was approximately 12 cm in diameter and stood 13.5 cm tall, as measured from the center of the fan to the table. The fan had three levels of varying speeds, 2 m/s, 3.1 m/s, and 3.6 m/s, respectively. These speeds were measured 20 cm away from the center of the fan using the averaging function built into a digital anemometer (brand: BTMETER BT-100 Handheld Anemometer, BTMETER,) over a minute long interval. The third level, 3.6 m/s, was used in this experiment, and the speed of the wind was checked between each trial when the bee was not present to ensure that approximately 3.6 m/s was achieved for each trial.

2.3. Procedures

Following well established protocols (e.g., [25–28]), the honey bees were randomly pulled while feeding from the 8% feeder table and captured in matchboxes. Once captured, the bees were brought to the experimental table and slowly released from the matchbox while feeding from the 50% sucrose (by volume) solution that was placed on the neutral gray training target. While feeding, the bees were marked on their abdomen, thorax, or both with "OPI" brand nail polish. Four colors (big apple red: #NLN25, mi casa es blue casa: #NLM92, exotic birds do not tweet: #NLF91, that's hula-rious: #NLH65) of OPI Nail Lacquer (OPI Products Inc., Calabasas, CA, USA) were used as up to four bees were simultaneously observed. Since many bees did not return to the table after their first visit during the two acquisition (i.e., shaping) trials, training multiple bees made data collection more efficient. Once the bee had been marked it was allowed to fly back to the hive and return to the neutral gray target two times before beginning the experimental phase of the study. During these two trials the fan was placed 20 cm to the right of (relative to the hive, see Figure 3) the gray training target but was not turned on. Once the bee completed the two shaping trials, the gray target was replaced by the orange target to begin the experimental phase which consisted of three control and nine experimental (12 total) trials. All trials were recorded with a GoPro Hero 10 camera (GoPro, Inc., San Mateo, CA, USA) set to 1080p, 60FPS, and a linear lens setting. A flex-arm camera mount held the camera in place, but no set position was used across all days of the experiment. Top (overhead) and side angles were used from which the three behavioral variables (see below) could easily be coded.

During these trials the fan was placed in three positions: at the front, to the left, or to the right of the orange target. During the three control trials the fan was kept off so that the bee could learn to interact with the orange target without the disturbance of the wind from the fan. In other words, the first three trials were similar to the shaping trials where the bee learned the orange target is the desired target upon which to land. The fan was turned on after these three trials were completed. The fan's starting position varied by group and stayed in the same position for three consecutive trials before being moved to the next position. In total there were six groups to control for the positionality of the fan (see Table 1). After the first three trials and based on their assigned grouping, each individual bee experienced three trials with the fan at the front, three trials with the fan at the left, and three trials with the fan on the right side of the landing site. At all times the fan was positioned 20 cm away from the center of the landing target.

Table 1. Fan positions.

Trial	Group					
	1	2	3	4	5	6
1	F	F	L	L	R	R
2	F	F	L	L	R	R
3	F	F	L	L	R	R
4	F	F	L	L	R	R
5	F	F	L	L	R	R
6	F	F	L	L	R	R
7	L	R	F	R	F	L
8	L	R	F	R	F	L
9	L	R	F	R	F	L
10	R	L	R	F	L	F
11	R	L	R	F	L	F
12	R	L	R	F	L	F

Note. During the first three trials (above the solid line) the fan was turned off and was placed in the same position as when the fan was turned on for trial four. The fan was placed in one of three positions (R = Right, F = Front, L = Left), as shown in Figure 3.

2.4. *Behavioral Coding and Dependent Variables*

There were three primary dependent variables in the current experiment: (1) landing successes, (2) number of approaches to the target, and (3) direction of approaches. The landing successes were recorded on site during the experiment whereas the approach observations were taken from video recordings of the trials. Video recordings were used to tally the number of approaches and how the bees approached the target. A successful landing was defined as the bee landing and feeding from the 25 mL drop of 50% sucrose solution. Accordingly, we expected one successful landing per trial, unless the bee landed, fed, and then lifted off and re-landed and fed because something interfered with the bee while it was feeding from the sucrose solution. It should be noted that the camera overheated on several occasions, resulting in the loss of data for several trials for four honey bees (cases #5, #8, #9, and #11).

The bees varied in the number of times and in the manner which they approached the target prior to landing and feeding. Upon reaching the table the bees would hover, approach the target, retreat, change position, and move toward the target again, sometimes from different directions. Observing this behavior on the videotapes, an approach was tallied when the bee's head was positioned towards the target (i.e., directed at it) while simultaneously flying in the wind stream. Approach direction was recorded with four possible outcomes: (1) headwind, (2) tailwind, (3) left, (4) right. Headwind meant the bee was approaching the target into the wind. Tailwind meant the bee was approaching the target with the wind. Left and right meant that the bee was approaching from either the left or right side of the table relative to the hive (see Figure 3), respectively.

Lastly, the intertrial intervals were also recorded from the moment the bee left the table to the moment it landed and received the reward. A limit of 20 min was placed on each bee to return to the feeding table. Only one bee (case #11) failed to return within this time period after the 11th trial. Data for this bee are nonetheless included in the analyses below since it had completed 92% of the total trials. Upon completion of the 12th trial, each bee was captured and sacrificed to ensure it was not reused and that every bee was experimentally naïve.

3. Results

3.1. Successful Landings

Having trained to fly to the table and feed upon the sucrose solution, could the bees overcome the added wind disturbance and still reach the target? To answer this question, we analyzed the recorded number of landing successes for each individual honey bee. Specifically, we tallied the number of successes across the three control trials (1–3) and separately across the experimental trials (4–12) and converted the results to percentages; viz., the Percent Correct Classifications (PCC) indices. A randomization test was then conducted for each analysis to determine a plausibility value, referred to as a chance value (*c*-value; see [29,30]), which aids in assessing whether or not the observed PCC index can be attributed to physical chance. Like traditional *p*-values, low *c*-values (e.g., <0.05) are desirable to support a causal interpretation of the outcome.

Table 2 presents the results from these analyses, and as can be seen, the PCCs for 13 out of 14 bees were 100% (*c*-values < 0.001) for both the control and experimental trials, therefore indicating they successfully landed on the target and fed from the sucrose reward on every trial, with or without the presence of wind. Aggregating across all trials and all bees, the bees were successful on 159 out of 167 trials; PCC = 95.21%, $c < 0.001$. Only one bee (case #11, see Table 2) was unable to consistently land on the target once the wind was turned on. This bee was able to land and feed during the first three trials, but once the wind was turned on it never successfully landed on the target. This bee also failed to return to the table on the 12th trial and was discontinued after 20 min as defined in the methods.

Table 2. Successful landing results.

Case #	Trials 1–3			Trials 4–12		
	Trials	PCC	c-Value	Trials	PCC	c-Value
1	3	100.00	0.123	9	100.00	0.002
2	3	100.00	0.126	9	100.00	0.002
3	3	100.00	0.124	9	100.00	0.002
4	3	100.00	0.124	9	100.00	0.002
5	3	100.00	0.125	9	100.00	0.003
6	3	100.00	0.126	9	100.00	0.001
7	3	100.00	0.125	9	100.00	0.002
8	3	100.00	0.124	9	100.00	0.001
9	3	100.00	0.124	9	100.00	0.002
10	3	100.00	0.125	9	100.00	0.002
11	3	100.00	0.126	8	0.00	1
12	3	100.00	0.125	9	100.00	0.002
13	3	100.00	0.126	9	100.00	0.002
14	3	100.00	0.125	9	100.00	0.001
Total	42	100.00	<0.001	125	93.60	<0.0001

Note. The PCC is the percentage of trials the bee was able to successfully land and feed from the sucrose reward. The *c*-value is a proportion from a randomization test, and low values indicate the observed PCC index is not plausibly the result of physical chance.

3.2. Approaches and Direction

Examination of the recorded videos revealed that the bees varied the rates (i.e., their speeds) at which they flew through the wind as well as the direction from which they approached the target. The bees could also be seen pushing against the wind, being pushed back, and then pushing forward again through the wind. These behaviors can readily be seen in a number of example videos posted in an online repository (Video Example X (https://drive.google.com/file/d/1zb-OJtuMbTlyamHSAtK6GlngumXqIwLa/view?pli=1, uploaded on 9 July 2025; last accessed on 15 August 2025) & Video Example Y (https://drive.google.com/file/d/1Fn6he857-UY2DHK6f-wK84G-xnsCmKQg/view, uploaded on 9 July 2025; last accessed on 15 August 2025). Given this variability, were the bees observed to demonstrate a general preference in the direction they approached the target? For descriptive purposes only, we first tallied the number of approaches across all bees for each of the four directions: into the wind (i.e., headwind), with the wind (i.e., tailwind), from the left side of the target (i.e., left), or from the right side of the target (i.e., right). As can be seen in Figure 4, across the first three trials in which the fan was turned off, the bees tended to reduce the number of approaches needed to land on the target. Importantly, while the tailwind approach was used less frequently than the others, no strong preference is visible in the plotted totals. When the fan was turned on for the remaining experimental trials, however, a clear preference for the headwind approach was observed across the nine experimental trials, while the tailwind was generally avoided. Similarly to when the fan was turned off, the bees also tended to require fewer approaches to successfully land on the target from the fourth to the twelfth trial. For the first experimental trial (Trial #4; see Figure 4), all fourteen honey bees used more headwind approaches than tailwind or right side approaches (PCCs = 100%, c's < 0.01). Eleven of the bees used more headwind approaches than left side approaches (PCC = 78.57%, $c = 0.01$). By the final trial (Trial #12), the majority of bees still used the headwind approach more frequently than the tailwind (PCC = 69.23, $c = 0.03$), right side (PCC = 76.92, $c = 0.01$), or left side (PCC = 76.92, $c < 0.01$), although the results were not as pronounced.

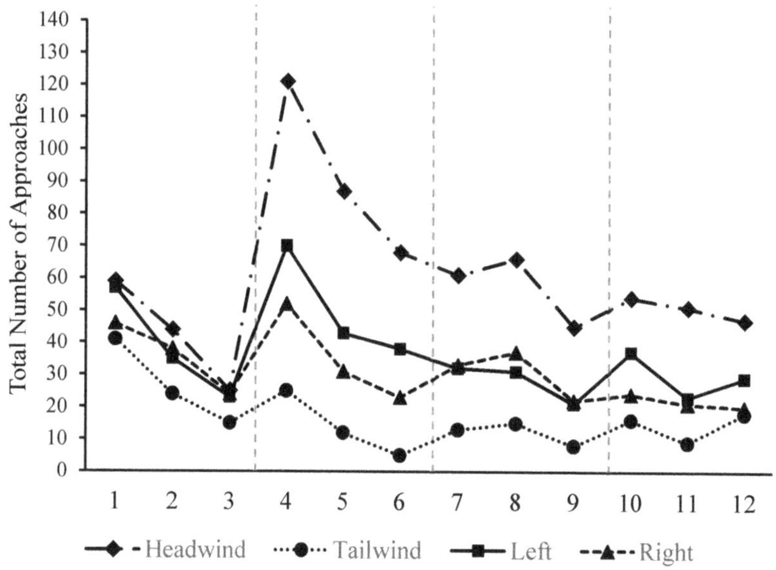

Figure 4. Total number of approaches across all bees and for four approach directions. Note. Dotted vertical lines demarcate changes in the fan's position. During Trials 1–3 the fan was off. During trials 4–12 the fan was on but was rotated to three positions (right, front, or left) every three trials (i.e., Trials 4–6, 7–9, and 10–12).

As a general test of the headwind preference across all twelve experimental trials shown in the aggregate tallies in Figure 4, we constructed a logic statement for the number of approaches. The logic statement was classified as 'true' if and only if the number of headwind approaches was greater than the number of tailwind approaches, the number of right side approaches, and the number of left side approaches for a given trial:

$$\text{True} \leftrightarrow (\text{H}_\text{k} > \text{T}_\text{k}) \wedge (\text{H}_\text{k} > \text{L}_\text{k}) \wedge (\text{H}_\text{k} > \text{R}_\text{k}),$$

where $\wedge$ represents logical conjunction, $_\text{k}$ represents the paired *k*th trial, and H, T, L, and R represent 'headwind,' 'tailwind,' 'left,' and 'right,' respectively. For example, the third bee (case #3) approached the target 16 times on trial #4: eight times into the headwind, two times with the tailwind, and three times each from the left or right directions. The logical statement, $(8 > 2) \wedge (8 > 3) \wedge (8 > 3)$, was therefore true for this particular bee and trial. Its number of approaches into the headwind was greater than its number of approaches for each of the other three directions. The number of times the logical statement was true across the three control or the nine experimental trials was tallied for each bee and reported as Percent Correct Classifications indices and evaluated with randomization tests.

Table 3 presents the results for the first three control trials (1–3) and the nine experimental trials (4–12). For the first three trials, since the fan was not on, the bees were not expected to show a clear directional preference. Results supported this expectation as only three bees (cases #1, #6, and #7) showed a slight preference (PCCs = 66.67%) for approaching the target while facing the fan. Aggregating across all of the bees, only 9 out of 41 trials were classified as 'true,' yielding a PCC equal to 21.95%, $c = 1.00$. During the first 3 trials, then, the bees did not prefer to fly towards the fan (which would be the headwind if the fan was on) more than the other approaches when landing on the target.

Table 3. Headwind preference results.

Case #	Trials 1–3			Trials 4–12		
	Trials	PCC	c-Value	Trials	PCC	c-Value
1	3	66.67	0.51	9	77.78	0.09
2	3	0	1	9	66.67	0.26
3	3	0	1	9	88.89	0.02
4	2	0	1	9	66.67	0.26
5	3	33.33	0.87	8	50.00	0.64
6	3	66.67	0.50	9	77.78	0.09
7	3	66.67	0.51	9	66.67	0.25
8	3	0	1	8	75.00	0.15
9	3	0	1	8	62.50	0.37
10	3	0	1	9	66.67	0.25
11	3	33.33	0.87	7	42.86	0.77
12	3	0	1	9	66.67	0.25
13	3	0	1	9	77.78	0.09
14	3	33.33	0.88	9	77.78	0.09
Total	41	21.95	1.00	121	69.42	<0.001

Note. The PCCs indicate percentages of trials the bees showed a preference for the headwind directional approach over the other three options. Some bees have fewer trials because the video was unavailable to review the number of approaches. For trials 1–3 the fan was off; for trials 4–12 the fan was on.

For trials 4 through 12, however, the bees' preferences changed. As can be seen in Table 3, six bees showed strong preferences for the headwind approach (PCCs $\geq$ 75%, c's $\leq$ 0.15), while another six bees showed moderate preferences for this approach (PCCs > 62%, c's $\leq$ 0.26). Only two bees (case #5 and case #11) failed to show a clear headwind preference (i.e., PCC $\leq$ 50%) in their approach preferences. Across all bees 84 out of 121 observations matched expectation (i.e., were true for the logic statement), yielding a PCC equal to 69.42%, $c < 0.0001$. These results reveal that when the wind was turned on, the bees approached the target into the wind (i.e., headwind) more than any other direction. It should also be noted that *c*-values, like traditional *p*-values, are influenced by sample size and the number of possible outcomes (in this case two: true vs. false). With a larger number of trials or more possible outcomes, the *c*-values for the individual PCCs magnitudes above would be smaller.

Finally, we evaluated the direction of the last approach each bee made before it successfully landed on and fed from the target. We specifically tallied the number of times the successful approach was from the headwind direction rather than from one of the other three directions. Results revealed PCCs equal to 100% (*c*-values < 0.0001) for ten of the fourteen bees; meaning, these bees successfully landed on the target by approaching into the wind (i.e., a headwind) across all of the trials that the wind was turned on (i.e., trials 4–12). Recall from above that one bee (case #11) made many attempts to land, but never successfully landed. Aggregating across all bees and trials when the fan was on, the bees landed successfully by using a headwind approach on 111 out of 114 trials; PCC = 97.37, $c < 0.0001$. There were only three instances across all trials and all bees during which the bees successfully landed using a direction other than the headwind approach. Cases #2 and # 14 each made one successful landing after approaching from the left and case #7 made a successful landing after approaching with the wind (i.e., tailwind), though video analysis revealed this to be a rather haphazard landing.

4. Discussion

Self-regulatory systems are best understood as structures and processes organized into non-linear, dynamic wholes. Well known examples include the positive feedback loop mechanism underlying blood clotting and the negative feedback loop mechanism underlying body temperature regulation [31–33]. Behavior, too, can be modeled in ways that go beyond the traditional linear causation, input–output approach. Drawing upon the tenets of Perceptual Control Theory we designed our study so that honey bees would have to dynamically modify their behavior to counteract a disturbance and reach a rewarding target. More specifically, the bees had to modify the speed of their flight and the direction of their flight paths to counteract wind disturbances from varying directions in order to land on a target and receive a sucrose reward. If behavior is regarded as the control of perception, then identifying and manipulating a controlled variable is an important part of experimentation (see also [2]). In this case the controlled variable was the distance between the bee's current position and the target, and the ideal distance was a value of '0' (i.e., no distance from the target). In accord with the negative feedback loop in Figure 1, the bees achieved this internal standard by varying the direction which they approached the target and the speed of their flight to overcome the 3.6 m/s wind and reach the target.

Results revealed the bees preferred to approach the target by flying into the wind as a means for counteracting the disturbance. The fan position was moved (i.e., disturbed) every three trials to a new position, but regardless of the position the bees still approached into the wind more than any other direction; meaning, the bees were not approaching the target consistently from the same direction (e.g., from the back side of the landing platform); but rather, they varied their approaches to match a preferred approach (viz., approach into wind). From the observational data, the bees were seen varying the speeds with which they were flying, pushing against the wind to get closer to the target, getting pushed back, and then pushing forward again and eventually landing. Across almost all the trials (97.37%), the bees approached the target into the wind (i.e., headwind) on their final approach before successfully landing and feeding on the target. It is important to point out that no two bees showed identical strategies in how they eventually approached the target facing into the headwind. Some bees tried various approaches from different directions and altitudes while others quickly oriented themselves into the headwind. In other words, while the bees varied with regard to their individual flight paths and speeds, nearly all overcame the disturbance and eventually reached the target through the headwind. The one honey bee (case #11) that never landed successfully on any of the experimental trials showed no preference for a headwind approach.

Individual variability in the specific actions of organisms as they strive toward their goals is common in studies of complex behaviors, such as foraging. Perceptual Control Theory makes room for this variability in a number of ways which can be explained by returning to and considering Figure 1 further. The model in the figure is necessarily simplified and is sufficient for explaining the patterns of results in this study. A more complete model, however, would include multiple levels of hierarchically controlled variables (see [2,5,6,9]). During flight the bees must also control their stability by varying their body position (i.e., pitch and yaw). Consistent with the fundamental laws of aerodynamics, by approaching into the wind the bees will have greater stability when landing, the same as an airplane approaching a runway. Similarly, it is easier to respond to and counteract harsh changes (e.g., a crosswind) when flying into the wind than it is in other directions. Finally, flying into the wind (whether landing or taking off) has a well-known impact on lift and drag, which are important variables for flight (see [34–36]). In any given moment, each bee is simultaneously controlling all of these variables (among others) in harmony and in the context of continuously changing environmental feedback. With such complexity, individual variability is to be expected.

Generally speaking, results from studies regarding the flying dynamics of honey bees comport well with PCT. For instance, when flying through tunnels, bees maintain equidistance from the walls by balancing perceived image speeds, controlling flight position to match a centered reference [37]. Flight speed is regulated by holding average image velocity constant, reducing collision risks by stabilizing optic flow perception. Most relevant to the present work is the finding that honey bees adjust their speed to land at a slow speed, thus avoiding a crash. During landing, bees attempt to keep the rate of expansion of the visual image constant, thus slowing down to near-zero velocity at touchdown [38] and showing another form of the control of perceptual inputs, consistent with PCT. Distance to food sources is gauged by integrating optic flow, maintaining a visually driven 'odometer' that is robust to wind and energy variations [39]. These behaviors rely on distinct movement-sensitive mechanisms, separate from the optomotor response, suggesting multiple control systems managing the perceptual variables for navigation (see [36] for a review of these self-regulation behaviors). In different terminology, these navigational feats can be construed as the operation of servomechanisms [40] which are entirely consistent with Powers' PCT.

Figure 1 also omits the fact that multiple disturbances are most certainly at play for any given bee and trial. We controlled the wind disturbance with a fan, but other disturbances impacted each bee as it left the hive and flew towards the target. For example, the bee may have needed to navigate around other bees to successfully leave the hive, which would change its flight path to some degree. Other bees may also have been in the flight path, particularly when leaving the hive. While the experiment was conducted on days without strong winds, the wind likely fluctuated between trials and even when the bee was in flight. The appearance of the table, landing site, and reward likely fluctuated as well from cloud cover and the movement of the sun. All of these potential disturbances had to be counteracted by the bee as it left the hive and flew to the table. Again, Perceptual Control Theory makes room for these disturbances which are not under the control of the experimenter [19]. If the organism is controlling its perceptions, it will reliably reach its goal although not exactly the same way. As William James pointed out over one hundred years ago, "fixed end. . .varying means!" (as quoted in [6], p. 86).

More generally, Perceptual Control Theory makes room for greater nuance in modeling behavior because it fits under the umbrella of Aristotelian philosophy. Iconic models, like the image in Figure 1, have played important roles in the history of scientific discovery [41–43]. One need only think of ball-and-stick models of molecular compounds or Watson and Crick's famous double-helix model of Deoxyribonucleic Acid. Such models are explanatory in nature as they encapsulate the causal structures of the natural systems being studied. For Aristotle, there are four fundamental causes, or explanatory factors, that may be incorporated into a scientific explanation. Material cause refers to the physical substance or matter of which something is made, for example, the marble of a statue. Formal cause refers to the design, structure, or nature that defines what a thing is, such as the shape and features that make a statue recognizable as that of a man. Efficient cause identifies the agent or process that brings something into being or changes it in some important way, such as the numerous strikes of the sculptor's hammer. Finally, final cause is the purpose or end goal for which something or some process exists, such as the sculptor's reason for creating the statue or the function of foraging in honey bee behavior. Nuance can be added to discussions of final cause by using the word "teleology" to describe human goals and purposes, and the word "teleonomy" to describe the functions and end states found in the physiology and behavior of organisms as well as in other natural systems [44–47]. In sum, the four causes–matter, form, agency, and purpose–offer a comprehensive and holistic way of explaining systems in nature.

Perceptual Control Theory is teleological or teleonomic because it explicitly incorporates final causes (purposes, goals, internal standards, end-states) into its modeling approach (see [2,3,48]). Every control system will incorporate at least one internal set-point which serves as its goal or natural end state. From the perspective of the observer, this set-point is central to understanding or explaining the system. Any mention of the function of the system, for instance, will entail its final cause (e.g., the function of the immune system is to ward off and fight infection). The function of the negative feedback loop considered in this study is to guide the bees through disturbances so they may reach a food source–the goal or end point. Guided flight to a specific target is part of a larger system, foraging behavior, the purpose of which is the well-being of the individual bee as well as the success and survival of its colony and species. In a way, the negative feedback loops incorporated in Perceptual Control Theory are organized around their final causes. It is as if the final cause holds the system together, and by systematically varying disturbances, as was performed in the current study, such causes can be identified and studied.

The word "system" implies not just final causes, however, but structures and processes organized into a coherent whole—a Gestalt. Honey bees are of a particular nature (formal cause) which entails foraging behavior. Not all organisms, of course, are foragers, and hence the current study relied upon the given nature of honey bees for its design. As noted above, pointing out what might seem obvious nonetheless enhances our explanation of the bees' successful navigation through the disturbances to reach the target. Material causes can also be invoked in discussions of the sucrose solution, a bee's proboscis, eye structure, neuroanatomy, etc., to deepen our understanding of its behavior. Lastly, efficient causes are part-and-parcel of negative feedback loops as can be seen in the arrows in Figure 1. The arrows represent time-ordered sequences of changes in the system; for example, the Input Function sends a signal to the Comparator which then sends a signal to the Output Function. In this way, we can see that simple input–output models (e.g., $S \rightarrow O \rightarrow R$) can be subsumed by more holistic, dynamic models rooted in Aristotle's philosophy of nature.

Approaching the study of behavior from this vantage point may lead to new ideas and to new ways of understanding existing ideas or theories. For example, 'effort' and 'motivation' have always been of interest to the neo-behaviorists (e.g., [49–51]). In developing and considering previous experiments from the vantage point of Perceptual Control Theory, a novel methodology to test 'effort' in honey bees could be employed. The fan in the current study could be systematically varied to find the point at which the organism will stop attempting to land on the target. The fan speed was kept at 3.1 m/s, but the speed could be intensified on each trial to determine the point at which a bee would no longer attempt to land and leave, so long as the maximum speed does not exceed its maximum flight speed capabilities. Such research would also benefit from more precise measurements of flight speeds and approach vectors by filming the bees from a set top position. With a circular area mapped onto the landing site, exact measurements could then be taken and additional navigational quantities computed (see [52] for a review of research focused on insect navigation). The reward could also be varied to determine if the bees are responding to a specific 'quality' of reward (see [53,54]). More generally, systematic variation and rigorous control procedures are methods which were developed and are routinely employed by comparative psychologists. Accordingly, Perceptual Control Theory may breathe new life into a field that is wrongly considered by many to be outdated [55].

In summary, scientists need not think that behaviors are a byproduct of blind, trial and error processes (see [5]); rather, they may be understood as teleonomic. . .as reaching ends with regularity. If Yin [3] is correct in suggesting that a ". . .restoration of purpose is therefore a new beginning for the sciences of behavior (p. 345)," then Perceptual Control Theory and other tools for modeling the dynamics of complex organic systems may lead to new ways of studying honey bee behavior. With an increase in our understanding of honey bee behavior, we scientists can develop more effective strategies to protect pollinators, enhance agricultural productivity, and mitigate threats like colony collapse disorder and environmental stressors.

Author Contributions: I.T.J. conducted the experiment, analyzed the data, and wrote most of the manuscript which is based on his dissertation [56]. J.W.G. wrote portions of the manuscript and assisted with data analysis. C.I.A. helped to design and conduct the experiment. All authors have read and agreed to the published version of the manuscript.

Funding: This research received no external funding.

Data Availability Statement: Data can be seen via the link: https://osf.io/4dt9p/ (accessed on 6 July 2025).

Conflicts of Interest: Ian T. Jones is employed by the Applied Research and Analysis Company, LLC. The remaining authors declare that the research was conducted in the absence of any commercial or financial relationships that could be construed as a potential conflict of interest.

References

1. Hall, J.E.; Hall, M.E. *Guyton and Hall Textbook of Medial Physiology*, 14th ed.; Elsevier: Amsterdam, The Netherlands, 2021.
2. Marken, R.S. *The Study of Living Control Systems: A Guide to Doing Research on Purpose*; Cambridge University Press: Cambridge, UK, 2021. [CrossRef]
3. Yin, H.H. Restoring purpose in behavior. In *Computational and Robotic Models of the Hierarchical Organization of Behavior*; Springer: Berlin/Heidelberg, Germany, 2013; pp. 319–347. [CrossRef]
4. Arocha, J.F. Scientific realism and the issue of variability in behavior. *Theory Psychol.* **2021**, *31*, 375–398. [CrossRef]
5. Powers, W.T. *Behavior: The Control of Perception*, 1st ed.; Aldine Publishing Company: Chicago, IL, USA, 1973.
6. Runkel, P.J. *Casting Nets and Testing Specimens: Two Grand Methods of Psychology*; Praeger Publishers: Westport, CT, USA, 2007.
7. Powers, W.T.; Clark, R.K.; McFarland, R.L. A general feedback theory of human behavior: Part I. *Percept. Mot. Ski.* **1960**, *11*, 71–88. [CrossRef]
8. Powers, W.T.; Clark, R.K.; McFarland, R.L. A general feedback theory of human behavior: Part II. *Percept. Mot. Ski.* **1960**, *11*, 309–323. [CrossRef]
9. Powers, W.T. Quantitative analysis of purposive systems: Some spadework at the foundations of scientific psychology. *Psychol. Rev.* **1978**, *85*, 417–435. [CrossRef]
10. Ashby, W.R. *An Introduction to Cybernetics*; Chapman & Hall: London, UK, 1956. [CrossRef]
11. Mead, M. Cybernetics of Cybernetics. In *American Society for Cybernetics, Purposive Systems, Proceedings of the First Annual Symposium of the American Society for Cybernetics, Gaithersburg, MD, USA, 26–27 October 1967*; Foerster, H.V., Ed.; Spartan Books: Washington, DC, USA, 1968.
12. Rosenblueth, A.; Wiener, N.; Bigelow, J. Behavior, purpose and teleology. *Philos. Sci.* **1943**, *10*, 18–24. [CrossRef]
13. Wiener, N. *Cybernetics*; John Wiley: New York, NY, USA, 1948.
14. Brembs, B. The brain as a dynamically active organ. *Biochem. Biophys. Res. Commun.* **2021**, *564*, 55–69. [CrossRef]
15. Marken, R.S. You say you had a revolution: Methodological foundations of closed-loop psychology. *Rev. Gen. Psychol.* **2009**, *13*, 137–145. [CrossRef]
16. Shaffer, D.M.; Marken, R.S.; Dolgov, I.; Maynor, A.B. Chasin'choppers: Using unpredictable trajectories to test theories of object interception. *Atten. Percept. Psychophys.* **2013**, *75*, 1496–1506. [CrossRef]
17. Bell, H.C.; Pellis, S.M. A cybernetic perspective on food protection in rats: Simple rules can generate complex and adaptable behavior. *Anim. Behav.* **2011**, *82*, 659–666. [CrossRef]
18. Bell, H.C. Control in Living Systems: An Exploration of the Cybernetic Properties of Interactive Behaviour. Doctoral Dissertation, Department of Neuroscience, University of Lethbridge, Lethbridge, AB, Canada, 2018.
19. Bell, H.C. Behavioral variability in the service of constancy. *Int. J. Comp. Psychol.* **2014**, *27*, 338–360. [CrossRef]
20. Powers, W.T. A feedback model for behavior: Analysis of a rat experiment. *Behav. Sci.* **1971**, *16*, 558–563. [CrossRef]
21. Hendriksma, H.P.; Hsiung, K.K.; Broccard-Bell, H.C.; Nieh, J.C. Aggression between invasive hymenopterans in Southern California: Scutellata-hybrids versus German yellowjackets. *J. Apic. Res.* **2023**, *64*, 147–149. [CrossRef]
22. Albert, D.J. What's on the mind of a jellyfish? A review of behavioural observations on Aurelia sp. jellyfish. *Neurosci. Biobehav. Rev.* **2011**, *35*, 474–482. [CrossRef] [PubMed]

23. Iwai, N.; Sugiura, S.; Chiba, S. Prey-tracking behavior in the invasive terrestrial planarian Platydemus manokwari (Platyhelminthes, Tricladida). *Naturwissenschaften* **2010**, *97*, 997–1002. [CrossRef]
24. Pellis, S.M.; Bell, H.C. Unraveling the dynamics of dyadic interactions: Perceptual control in animal contests. In *The Interdisciplinary Handbook of Perceptual Control Theory*; Academic Press: Cambridge, MA, USA, 2020; pp. 75–99. [CrossRef]
25. Menzel, R.; Erber, J. Learning and memory in bees. *Sci. Am.* **1978**, *239*, 102–110. [CrossRef]
26. Mujagic, S.; Erber, J. Sucrose acceptance and different forms of associative learning of the honey bee (*Apis mellifera* L.) in the field and laboratory. *Front. Behav. Neurosci.* **2010**, *4*, 46. [CrossRef]
27. Rodriguez, S.D.; Wincheski, R.J.; Jones, I.T.; Jesus-Soto, D.; Michael, G.; Fletcher, S.J.; Pretends Eagle, T.J.; Grice, J.W.; Abramson, C.I. Some phenomena of the cap-pushing response in honey bees (*Apis mellifera* spp.). *J. Comp. Psychol.* **2023**, *137*, 178. [CrossRef]
28. Wincheski, R.; Jones, I.T.; Rodriguez, S.; Soto, M.; Fletcher, S.; Eagle, T.; Grice, J.; Abramson, C. Training Honey Bees (*Apis mellifera*) to Push a Cap: Shaping, Observational Learning, and Memory. *Int. J. Comp. Psychol.* **2023**, *36*, 1–10. [CrossRef]
29. Grice, J.W. *Observation Oriented Modeling: Analysis of Cause in the Behavioral Sciences*; Academic Press: Cambridge, MA, USA, 2011.
30. Grice, J.W. Drawing inferences from randomization tests. *Personal. Individ. Differ.* **2021**, *179*, 110931. [CrossRef]
31. Duan, B.; Xu, X.S. How to break a fever: A feedback circuit for body temperature control. *Neuron* **2019**, *103*, 179–181. [CrossRef]
32. Mitchell, D.; Fuller, A.; Snelling, E.P.; Tattersall, G.J.; Hetem, R.S.; Maloney, S.K. Revisiting concepts of thermal physiology: Understanding negative feedback and set-point in mammals, birds, and lizards. *Biol. Rev.* **2025**, *100*, 1317–1346. [CrossRef]
33. Tan, C.L.; Knight, Z.A. Regulation of body temperature by the nervous system. *Neuron* **2018**, *98*, 31–48. [CrossRef] [PubMed]
34. Barron, A.; Srinivasan, M.V. Visual regulation of ground speed and headwind compensation in freely flying honey bees (*Apis mellifera* L.). *J. Exp. Biol.* **2006**, *209*, 978–984. [CrossRef] [PubMed]
35. Nachtigall, W.; Rothe, U.; Feller, P.; Jungmann, R. Flight of the honey bee: III. Flight metabolic power calculated from gas analysis, thermoregulation and fuel consumption. *J. Comp. Physiol. B* **1989**, *158*, 729–737. [CrossRef]
36. Srinivasan, M.V.; Zhang, S.W.; Lehrer, M.; Collett, T.S. Honey bee navigation *en route* to the goal: Visual flight control and odometry. *J. Exp. Biol.* **1996**, *199*, 237–244. [CrossRef]
37. Baird, E.; Srinivasan, M.V.; Zhang, S.; Cowling, A. Visual control of flight speed in honeybees. *J. Exp. Biol.* **2005**, *208*, 3895–3905. [CrossRef]
38. Baird, E.; Boeddeker, N.; Ibbotson, M.R.; Srinivasan, M.V. A universal strategy for visually guided landing. *Proc. Natl. Acad. Sci. USA* **2013**, *110*, 18686–18691. [CrossRef]
39. Srinivasan, M.V.; Zhang, S.; Altwein, M.; Tautz, J. Honeybee navigation: Nature and calibration of the "odometer". *Science* **2000**, *287*, 851–853; Erratum in *Science* **2000**, *287*, 1756. [CrossRef] [PubMed]
40. Cheng, K. Oscillators and servomechanisms in orientation and navigation, and sometimes in cognition. *Proc. R. Soc. B* **2022**, *289*, 20220237. [CrossRef]
41. Frigg, R. *Models and Theories, a Philosophical Inquiry*; Routledge: Abingdon, UK, 2023. [CrossRef]
42. Grice, J.W.; Jones, I.T. Iconic models in science and psychology. *Theory Psychol.* **2024**, *34*, 815–835. [CrossRef]
43. Harré, R. *Modeling: Gateway to the Unknown Studies in Multidisciplinarity*; Rothbart, D., Ed.; Elselvier: Amsterdam, The Netherlands, 2004; Volume 1. [CrossRef]
44. Ayala, F.J. Teleological explanations in evolutionary biology. *Philos. Sci.* **1970**, *37*, 1–15. [CrossRef]
45. Mayr, E. The idea of teleology. *J. Hist. Ideas* **1992**, *53*, 117–135. [CrossRef]
46. Wright, L. *Teleological Explanations: An Etiological Analysis of Goals and Functions*; University of California Press: Berkeley, CA, USA, 1976. [CrossRef]
47. Wallace, W.A. *The Modeling of Nature: Philosophy of Science and Philosophy of Nature in Synthesis*; The Catholic University of America Press: Washington, DC, USA, 1996.
48. Marken, R.S. The dancer and the dance: Methods in the study of living control systems. *Psychol. Methods* **1997**, *2*, 436–446. [CrossRef]
49. Tolman, E.C. A cognition motivation model. *Psychol. Rev.* **1952**, *59*, 389. [CrossRef]
50. Hull, C.L. *Principles of Behavior: An Introduction to Behavior Theory*; Appleton-Century: New York, NY, USA, 1943.
51. Spence, K.W.; Spence, J.T. *Psychology of Learning and Motivation*; Academic Press: Cambridge, MA, USA, 1967; Volume 1.
52. Buehlmann, C.; Mangan, M.; Graham, P. Multimodal interactions in insect navigation. *Anim. Cogn.* **2020**, *23*, 1129–1141. [CrossRef] [PubMed]
53. Gil, M.; De Marco, R.J.; Menzel, R. Learning reward expectations in honey bees. *Learn. Mem.* **2007**, *14*, 491–496. [CrossRef] [PubMed]
54. Sanderson, C.E.; Cook, P.; Hill, P.S.; Orozco, B.S.; Abramson, C.I.; Wells, H. Nectar quality perception by honey bees (*Apis mellifera ligustica*). *J. Comp. Psychol.* **2013**, *127*, 341. [CrossRef] [PubMed]

55. Abramson, C.I. Why the study of comparative psychology is important to neuroscientists. *Front. Behav. Neurosci.* **2023**, *16*, 1095033. [CrossRef]
56. Jones, I.T. On The Origins of Behavior: Restoring Teleological Accounts of Behavior. Doctoral Dissertation, Oklahoma State University, Stillwater, OK, USA, 1875. Available online:https://www.proquest.com/openview/e0756339d6c32f99b9b855df57002139/1? (accessed on 6 July 2025).

insects

MDPI

Article

Evaluation of a Simple Antibiotic-Free Cryopreservation Protocol for Drone Semen

Sophie Egyptien [1,*], Jérôme Ponthier [1], Fabien Ectors [2], Brice Thibaut [3] and Stéfan Deleuze [1]

1 Fundamental and Applied Research for Animals and Health Research Unit (FARAH), Comparative Veterinary Medicine, Faculty of Veterinary Medicine, University of Liège, 4000 Liège, Belgium
2 Fundamental and Applied Research for Animals and Health Research Unit (FARAH), Veterinary Public Health, Faculty of Veterinary Medicine, University of Liège, 4000 Liège, Belgium
3 Independent Researcher, 4000 Liège, Belgium
* Correspondence: segyptien@uliege.be; Tel.: +32-43664200

Simple Summary: Modern agriculture relies mainly on the honey bee for pollination. Therefore, it is crucial to preserve and increase bee populations. Breeding programs have a crucial role, and insemination with frozen semen bypasses many practical limitations. This study evaluates the effectiveness of a simple antibiotic-free freezing protocol by assessing semen viability using two different tools and female brood production. The drop in viability associated with cryopreservation is 37%, and both evaluation tools give similar results. Despite that loss, 5/8 queens inseminated with frozen–thawed semen produced female brood, which is similar to the results obtained with the control group inseminated with fresh semen. This demonstrates that cryopreserved semen following this protocol is capable of egg fertilization. That, in turn, is very promising as this is the first antibiotic-free protocol reported in the literature to prove effective in female brood production.

Abstract: The increasing reliance of modern agriculture on honey bee (*Apis mellifera*) pollination has driven efforts to preserve and enhance bee populations. The cryopreservation of drone semen presents a promising solution for preserving genetic diversity and supporting breeding programs without live animal transport risks. This study aimed to evaluate a one-step dilution antibiotic-free drone semen slow-freezing protocol under field conditions with in vitro and in vivo parameters. Semen viability was tested by two different mixes of dyes, and both techniques gave similar results, showing a post-thaw viability drop of 37%. Virgin queens were inseminated either with fresh or frozen–thawed semen. Survival rates until egg-laying onset and female brood production were similar for both groups; however, colonies with queens inseminated with fresh semen were more likely to go into wintering. Results suggest that frozen–thawed semen can support viable queen insemination, with potential for female brood production even without antibiotics in the diluent. This study highlights the need for further refinement of cryopreservation protocols, particularly regarding semen quality and queen longevity, to improve the feasibility of cryobanking for *Apis mellifera* conservation and breeding.

Keywords: honey bee; drone semen; cryopreservation; antibiotic; viability; *Apis mellifera*; instrumental insemination

1. Introduction

The agricultural industry has become highly dependent on pollinators in recent decades, with *Apis mellifera*, the domesticated honey bee, playing a central role in crop pollination [1].

https://doi.org/10.3390/insects16010050

In many places, the growing demands of agriculture are no longer met [1] because of an insufficient global increase in the number of colonies and localized declines in populations, particularly in North America and several European countries [2].

Given their crucial role in pollination, it is essential to develop and improve methods for preserving the domesticated honey bees, along with subspecies and lineages adapted to diseases and predators. Preserving their genetic material can be achieved by maintaining live colonies or conserving live cells. Semen from drones, the haploid males, represent valuable cells to cryopreserve as a genetic safeguard.

Cryobanking of drone semen contributes to the preservation and restoration of colonies from known lineages, thus maintaining biodiversity. It supports breeding programs by allowing the shipping of semen worldwide, as the exchange of cryopreserved semen has proven to be a safe and effective [3] alternative to the transportation of live animals, which may be complex and often illegal at an international scale. Although proven effective, the transport of live material must always be carried out with caution as the container and the semen might represent vectors for infectious agents, such as bacteria or viruses, like in other species [4,5]. Cryopreserved semen can also be used for inseminations during periods when fresh semen is unavailable.

Slow-freezing protocols yielding satisfactory results have been reported on different occasions. Most of these involved several steps, including one or two dilutions, pre- or post-freezing centrifugation, and/or post-thaw cryoprotectant removal [6–12], with only a few [13–18] being single-step procedures. Some of the reports have evaluated post-thaw semen quality by in vitro techniques only [10–12,15,16,18], while others ran fertility trials to further confirm post-thaw quality [6–9,13,14]. All of the published protocols used antimicrobial agents. In the context of antimicrobial resistance, the non-therapeutic use of antibiotics, like in semen extenders, has become a matter of concern in many domestic species [19]. To the best of our knowledge, the production of female brood after insemination with an antibiotic-free frozen thawed drone semen antibiotic-free protocol has yet to be reported.

The most common semen quality parameter evaluated is viability. It is defined by the permeability of the plasma membrane: viable cells have an intact, impermeable plasma membrane [20]. Viability assessment is mostly performed under fluorescent microscopy using DNA markers such as propidium iodide (PI), which marks non-viable cells in red associated with SYBR-14 (SYBR) or Hoechst 33342 (Hoechst), both able to cross cell membranes regardless of their integrity [21]. Recently, Calcein Violet 450 AM (CaV), a dye marking the cytoplasm of cells with an intact membrane, has been suggested as a useful alternative to SYBR and Hoechst in several species, including drones [22].

The aim of this study is to evaluate a simple (one step dilution) antibiotic free protocol to freeze drone semen in vitro by comparing two combinations of dyes to assess post-thaw semen viability and in vivo by inseminating queens in a fertility trial.

2. Materials and Methods

The protocol was conducted in triplicate during the spring and summer of 2024.

2.1. Drone Semen Collection

Semen was collected once a month from May to July 2024 at the apiary using a Harbo syringe (Harbo, 1985) 9 to 16 days before freezing or insemination. Drones were collected until the glass straw was full (±150 µL). The glass capillary tube was then sealed and kept in the dark at 16 °C as described in standard procedures [8,12]. The three replicates of semen collection are referred to as 1, 2, or 3 hereafter.

2.2. Semen Freezing and Thawing

The day the semen was frozen, the semen sample was homogenized, 50 µL was loaded into a new sterile glass straw for insemination planned the next day, and 50 µL was prepared for freezing.

The dilution and straw-filling process were adapted from a previously published protocol [8]. Briefly, 50 µL of fresh semen was mixed with 33 µL of a cryopreservation diluent, referred to as Harbo's diluent, which consisted of 500 µL of buffer (NaH_2PO_4 and Na_2HPO_4, pH 7.21, ThermoFisher, Waltham, MA, USA), 250 µL of fresh egg yolk, and 250 µL of DMSO (ThermoFisher, Waltham, MA, USA). A 250 µL plastic straw was then filled with layers as follows: 20 µL of cryoprotectant, an air bubble, the diluted semen, another air bubble, and, finally, another 20 µL of cryoprotectant. One end was sealed with cotton and gelatin, while the other was closed using a manual ultrasonic sealer (CryoSealer, Minitüb GmbH, Tiefenbach, Germany).

The straw was then placed in a controlled-rate liquid nitrogen freezer (Freeze Control® Cryopreservation System, Cryologic Pty. Ltd., Victoria, Australia) with the following freezing protocol: starting at room temperature, the temperature was lowered at a rate of −1 °C per minute to 5 °C, held for 1 min, then reduced at −2 °C per minute to −6 °C, held for 1 min with ice seeding (to activate crystallization), and then at −5 °C per minute to −40 °C, before finally being plunged into liquid nitrogen and stored. The next day, semen was prepared for insemination and thawed in water at 35 °C for 30 s. Semen was expelled into a PCR tube, a sample of 2 µL was used for post-thaw semen analysis, and the rest was mounted in a new sterile glass capillary for insemination.

2.3. Instrumental Insemination

Twenty-nine queens, aged 7 to 12 days, were anesthetized with CO_2 the day before insemination for 3 min to stimulate oviposition and were inseminated following a standard method [8]. Insemination was conducted under CO_2 anesthesia and completed within 5 min of induction. A Pierre Marin insemination syringe (Pierre Marin, Forchies-la-Marche, Belgium) was used under a microscope to deposit 5 µL of either fresh (F) or frozen–thawed (Fz) semen in 13 and 16 queens, respectively. Three insemination sessions, each including both F and Fz semen, took place one month apart from May to July 2024, corresponding to the replicates of semen collection 1, 2, or 3.

Immediately after insemination, the queens were caged and placed in queenless nucleus hives with young bees (Mini-Plus hive, Stehr, Schwalmtal, Germany), where they remained for the duration of the experiment.

2.4. Brood Evaluation and Third Narcosis

Five days after introducing the queens into the hives, each hive was examined to confirm queen acceptance (release from her cage by worker bees) and the onset of egg laying. If a queen was accepted but had not begun egg laying within 11 days of insemination, she was subjected to a second round of CO_2 treatment to stimulate egg production [8]. Capped brood evaluations were performed every 3 to 5 days starting one week after egg observation, monitoring for the presence of female larvae.

2.5. Semen Analysis

When insemination doses were prepared, semen analysis was performed by the same operator for consistency, with fresh and frozen–thawed samples diluted at 1:200 in phosphate-buffered saline (PBS) before assessment.

2.5.1. Semen Concentration

Concentration was measured using a hemocytometer (Thoma cell counting chamber, Marienfeld®, Lauda-Königshofen, Germany). A 10 μL diluted semen sample diluted in 390 μL of 2% formaldehyde was used for counting. Spermatozoa counts from 10 squares of 0.2 mm were obtained under an optical microscope at 400× magnification. The concentration of undiluted semen was calculated using the formula: (number of sperm heads/0.04) × 40 × 200 = concentration in millions of spermatozoa/μL.

2.5.2. Motility

A 3 μL drop of diluted semen was placed in a pre-warmed slide (Leja, IMV technologies, Saint-Ouen-sur-Iton, France), and motility was grossly assessed after 5 min using brightfield optical microscopy. The microscope stage was kept at 37 °C, and motility was noted if any movement was observed.

2.5.3. Viability

Viability was assessed using fluorescent microscopy, and dyes were used following manufacturer recommendations and a previously described protocol [22]. The first fluorescent dyes mix (Mix 1) consisted of SYBR14 (SYBR) and propidium iodide (PI) from the LIVE/DEAD™ Sperm Viability kit, with the addition of Hoechst 33342 (Hoechst), which binds to the DNA of all cells and fluoresces in blue. The Hoechst stock solution was diluted 1:99 before staining 50 μL of the semen sample with 1 μL. When associated with PI, SYBR stains live sperm cells in green, and PI highlights dead cells in orange-red. The second mix of dyes (Mix 2) was composed of Draq5™ (Draq5), Calcein Violet 450 AM (CaV), and PI. Draq5 is a far red dye with the same properties as Hoechst. All dyes were sourced from ThermoFisher (Waltham, MA, USA). Images of spermatozoa were captured with an Observer 7 widefield fluorescent microscope (Zeiss, Oberkochen, Germany) equipped with a 40×/1.1 NA oil immersion objective. The samples were illuminated using wavelengths of 385 nm, 385 nm, 475 nm, 555 nm, and 653 nm for Hoechst, CaV, SYBR, PI, and Draq5, respectively. Brightfield imaging was also used. Emission wavelengths were filtered as follows: 430–470 nm for Hoechst and CaV, 500–550 nm for SYBR, 570–640 nm for PI, and 665–715 nm for Draq5. Images were automatically overlayed using Zen Blue 2.6 software (Zeiss, Oberkochen, Germany). The initial focus was achieved using the 430–470 nm emission filter in Mix 1 and 665–715 nm in Mix 2.

All samples were included in the analysis, regardless of semen quality and potential fertility.

3. Results

3.1. Semen Analysis

Semen analysis showed variability across replicates. Two of the three batches exhibited reduced viability after thawing. Viability results varied with the mix of dyes used. Mix 2 depicted lower viability when the number of cells evaluated was lower than 100. No moribund cell (SYBR and PI positive) was spotted with Mix 1 compared to Mix 2 where the percentage of moribund (CaV and PI positive) cells varied from 0 to 4% (F1: 2%; Fz1: 4%; F2: 2%; Fz2: 3%; F3: 0%; Fz3: 4%). Semen concentration in fresh samples ranged from 6.4 to 15.4 million sperm/μL, whereas frozen–thawed samples ranged from 4 to 6 million sperm/μL, affecting the number of sperm inseminated per queen. When technical issues allowed (Mix 1), the mean viability was 85% and 48% for F and Fz semen, respectively. Table 1 details the semen analysis results, including viability, concentration, total sperm cells per queen, the number of live sperm cells inseminated, and motility. Figure 1 illustrates sperm cells stained with Mix 1 and Mix 2.

Table 1. Semen analysis.

	Viability Mix 1 (%)	Viability Mix 2 (%)	Concentration (10^6/µL)	Sperm Cells/Queen (10^6)	Live Sperm Cells/Queen (10^6)	Motility (%)
F 1	75 (184/246) *	76 (86/113)	6.4	32	24	15
Fz 1	77 (179/232)	56 (25/45)	4	20	15	7
F 2	86 (168/195)	87 (113/130)	9.4	47	40	5
Fz 2	39 (88/225)	33 (33/101)	4.2	21	8	1
F 3	95 (140/148)	79 (26/33)	15.4	77	73	2
Fz 3	29 (44/154)	10 (10/99)	6	30	9	1

* (Live sperm cells/total number of sperm cells evaluated).

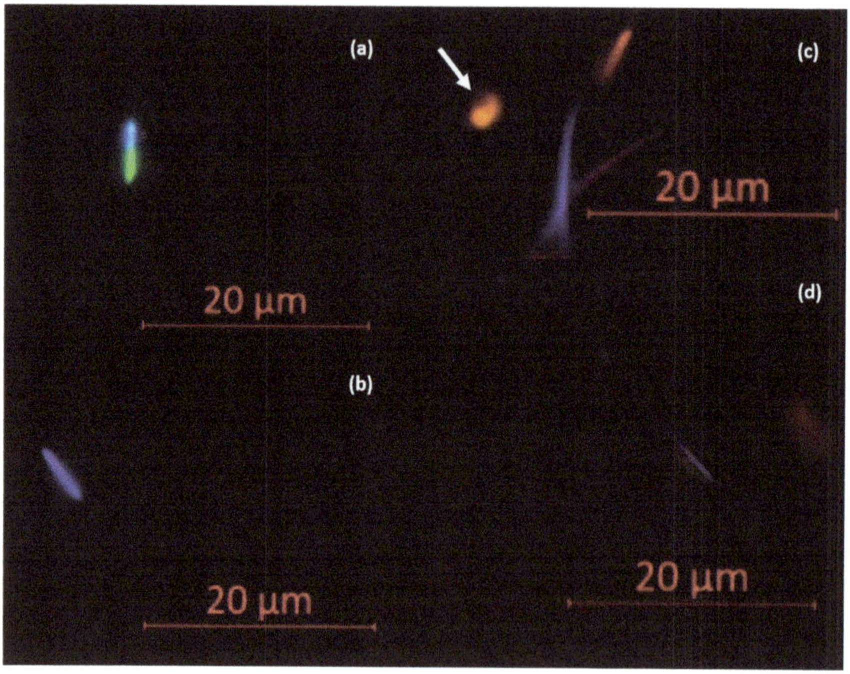

Figure 1. Fluorescence microscopy of sperm stained with Mix 1: Hoechst, SYBR, and PI (**a**,**b**), and Mix 2: Draq5, CaV, and PI (**c**,**d**). (**a**) The sperm cell is viable. The nucleus is stained with Hoechst (blue) and SYBR (green), indicating an intact membrane. (**b**) The sperm cell is non-viable. The nucleus is stained with Hoechst (blue) and PI (orange), indicating a damaged cell membrane. (**c**) The sperm cell marked by the arrow is non-viable. The nucleus is stained with Draq5 (red) and PI (orange). The second sperm cell is viable. The nucleus is stained with Draq5 and the cytoplasm with CaV (blue), indicating an intact cell membrane. (**d**) The sperm cell is moribund, its nucleus is stained with Draq5 (red) and PI (orange), and its cytoplasm is stained with CaV (blue).

3.2. Insemination Outcomes

The post-insemination average weight of the queens was 169.41 ± 0.02 mg. The queens' survival and acceptance rates were similar between groups. Some queens did not survive release, were not accepted by their colonies, or failed to lay eggs. Among surviving queens that laid eggs (five F and eight Fz), three from the F group and five from the Fz group produced female brood. One queen from the Fz1 group, after initially producing female brood, was quickly replaced by worker bees. The new queen mated naturally and went on to produce a typical proportion of female brood. Outcomes were similar in both groups in terms of survival until egg laying and female brood production.

Subjectively, the three F colonies and one from the Fz group were considered strong enough for overwintering, although at the time of the last examination before winter, the Fz was not as strong as the F colonies. Detailed insemination outcome data are provided in Table 2.

Table 2. Insemination results.

	Number of Queens That Survived Until:					
	Insemination	Release	Acceptation	Egg Laying	Female Brood	Overwintering Preparation
F 1	5	3	3	1	1	1
Fz 1	7	6	4	4	4	0
F 2	5	4	2	2	1	1
Fz 2	5	3	3	2	1	1
F 3	3	3	2	2	1	1
Fz 3	4	3	2	2	0	0
Total F	13	10	7	5	3	3
Total Fz	16	12	9	8	5	1

4. Discussion

Except for one report [14], female brood production following inseminations with frozen–thawed semen is less effective than after insemination with fresh semen. It is commonly accepted that the primary objective of insemination with frozen–thawed semen is to produce queens capable of generating sufficient fertilized eggs to establish new, healthy queens, thereby preserving genetic lines and supporting selective breeding [6]. It may be considered that the female brood produced from Fz was of good quality as one colony replaced the queen by rearing a new one from her fertilized egg. This new queen was naturally mated and thrived normally until the colony was robbed out six weeks later.

There are few studies on in vivo outcomes of inseminations using frozen–thawed semen and even fewer examining brood characteristics [6–9,23] despite this being one of the most accessible and essential field parameters. One of these studies found that queens inseminated with frozen–thawed semen laid fertilized eggs for approximately six weeks, a significantly shorter duration compared to queens inseminated with fresh semen, which continued laying throughout the season [8]. One of our Fz colonies successfully entered wintering and, as of mid-October, remained productive, though it was weaker than the three F colonies also prepared for overwintering. Overall, 1 out of 5 Fz colonies entered wintering, consistent with a previous study where 9 out of 41 colonies were up for wintering [9]. A recent study reported that 13 out of 33 queens inseminated with frozen–thawed semen survived and produced female brood for over three months, with all 13 colonies successfully preparing for wintering [14]. It demonstrates the feasibility of obtaining strong colonies inseminated with frozen semen but also illustrates the room left for improvement in insemination with cryopreserved drone semen.

The limited egg-laying period for queens inseminated with frozen–thawed semen warrants further study. The proportion of colonies that successfully overwinter could serve as a valuable in vivo metric for assessing improvements in freezing protocols, as it likely reflects the number of sperm reaching the spermatheca and their long-term survival. It has been suggested that when less than 0.5 million sperms reach the spermatheca, queens face an increased risk of failing to lay fertilized eggs [12]. In our experimental design, the total inseminated sperm count in the Fz group was lower than in the F group. In order to preserve sperm motility and viability, semen was not centrifuged after thawing and reconcentrated [7]. Centrifugation, even at 250 RCF for 10 to 20 min, can damage sperm membranes [24]. Consequently, the fertility of queens from the Fz group may be affected both immediately and in the long term, likely shortening their egg-laying period. Cryoprotectant toxicity might also have impacted their female brood production.

Cryoprotectants, even though necessary to avoid the formation of ice crystals during slow freezing, show toxicity for the semen and the queen itself [23], and their post-thaw removal has been encouraged [21]. This might be performed by centrifugation or dialysis after thawing [9]. Alternatively, reducing the toxicity of cryoprotectants has been achieved by pre-freezing dialysis [10]. Unfortunately, these protocols add procedural complexity. Therefore, we opted against incorporating them to maintain protocol simplicity. The most commonly used cryoprotectant for drone semen freezing is DMSO, whose optimal concentration balancing both toxicity and protective effects is 10% [6,10,15,21,23], which was used in this study. A recent publication reported promising results with a higher final concentration of 12% cryoprotectants, a combination of polyvinylpyrrolidone (PVP) or Dextran, and DMSO. This formulation improved post-thaw motility and membrane integrity, as well as overwintering success, with 3 out of 13 queens still producing female brood 10 months after insemination [14]. This suggests that cryoprotectant removal might not be as necessary as previously proposed [23] and that higher concentrations might improve in vivo results. Additionally, a higher diluent-to-semen ratio has been used in other studies with promising results [11,14]. Combining this approach with increased dilution ratios to our simple protocol might also improve post-thaw quality and should be investigated.

In our protocol, the semen diluent composition shows a limited number of components to keep it as simple as possible, and no catalase was added. Catalase has been shown to improve post-thaw viability and maintain post-thaw motility when compared to other catalase-free diluents [11]. The positive impact of the adjunction of catalase on in vitro parameters has yet to be confirmed in fertility tests.

No antibiotics were added to the semen diluent, marking this as the first report of female offspring production from frozen–thawed semen without any antimicrobial addition. Research has confirmed that drone semen is not sterile [25], and antibiotics are often used to prevent bacterial growth, as a high bacterial load negatively impacts semen quality [26]. However, some antimicrobial properties may help manage Nosema apis, a sexually transmitted parasite [25]. Additionally, the use of antibiotics raises concerns about fostering antimicrobial resistance and potential impacts on queen health [19], so any inclusion should be carefully considered. The chosen semen collection method (aspiration via glass pipette) minimizes contamination compared to seminal vesicle dissection, which has shown a 100% contamination rate in one study. Although bacterial load did not impact in vitro sperm viability or motility in that study [25], it may have influenced queen survival in our experiment.

The queen mortality rate was high but comparable between the Fz and F groups, suggesting a common stress on the colonies. During the 2024 Belgian breeding season, beekeepers experienced high queen mortality following insemination, likely linked to the queens' abnormally low weights. In our study, the average weight of the queens was 169.41 mg, whereas it is commonly accepted that a high-quality queen should weigh more than 200 mg after insemination. Queens with low weights, like those in our study, have been reported to experience lower acceptance rates and exhibit reduced sperm quantities in the spermatheca [27]. Semen quantity in the spermatheca is linked to the spermathecal volume, which is also linked to the weight of the queen [28]. The weight of a queen post-insemination is correlated with her weight at emergence [29], which is directly influenced by her rearing conditions, i.e., the age of the grafted larvae influences the weight at emergence [30]. Globally, queen-rearing success is influenced by available resources and environmental factors, such as humidity and temperature [31]. As all of these parameters were common to all F and Fz queens, it would have been interesting to multiply the number of queen breeders and test the effect of the queen-rearing impact on their survival. Additional environmental factors, such as ambient temperature, likely played a role, as three queens from the June batch appeared to die from overheating. At that time, we experienced a heat wave, and all of the bees nursing the queen were also found dead or dying. The other nuclei were moved to a cooler place before queen loss. This highlights the limitations of relying solely on in vivo assessments and highlights the need for reliable in vitro fertility metrics.

The highest post-thaw viability was observed in the first batch, which also exhibited the highest percentage of queens laying fertilized eggs. However, these queens did not sustain long-term female brood production and died within three weeks. Over 20 years ago, it was established that for a fixed inseminated volume of 8 μL of semen diluted 1:1, viability should reach at least 42% to produce only female brood, although female brood production has also been achieved with as low as 32% viable inseminated semen [32]. Unfortunately, in that study, semen concentration was not available, so the actual number of live sperm inseminated cannot be calculated. A fair estimation would be that around 40 million sperm were inseminated. This means that around 17 million live sperm cells would be necessary for producing exclusively female brood. Our results, partly agreeing with these estimations, suggest that it would be interesting to investigate the number of live sperm cells necessary to successfully produce female brood.

Overall, both methods used for viability assessment (CaV + PI and SYBR + PI) concurred. The main discrepancies were observed when less than 100 sperm cells could be evaluated. It is accepted that around 200 sperm cells should yield a precise evaluation of viability and that 100 cells should be a minimum [20]. Globally, the focus was more difficult to achieve when CaV was used in the fluorescent mix of dyes, rendering sample evaluation less effective for the same number of cells per microscopic field. To compensate, the number of acquired frames should have been increased. This could not be carried out as it prolongs light stimulation and thus increases photobleaching. As previously suggested, CaV associated with PI is efficient for evaluating viability but should most probably be used with flow cytometry instead of epifluorescence to take better advantage of this dye association. However, specific adjustments for drone semen and validation of the technique, with visual control under a microscope, are mandatory before the interpretation of results.

Overall, semen motility was low, characterized by minimal circular or active movement, with most sperm exhibiting linear position and vibratory motion. Various technical factors, such as chamber design and dilution parameters, can influence motility assessments, meaning results may vary under different conditions [33]. Recently, the CASABee software was developed to automatically analyze semen motility and concentration, thereby enhancing analytical accuracy [34]. Motility seems to be the best in vitro predictor of in vivo outcomes, such as sperm migration to the spermatheca and egg-hatching rates [6], although it may not correlate with overwintering success [14]. Automatic motility evaluation can be conducted in the field, requiring minimal technical expertise. Therefore, comparing manual observations with automatic measurements and in vivo metrics could have been useful in confirming motility as an accurate in vitro fertility predictor.

A primary limitation of this study is the low number of queens that initiated oviposition. Although the number of inseminated queens is comparable to those reported in other published protocols [7–9,14], statistical analyses were not conducted due to the limited number of queens that survived until oviposition. Nonetheless, the main objective of this study—testing the feasibility of obtaining female brood using a one-step antibiotic-free freezing protocol—has been successfully achieved.

5. Conclusions

This study highlights significant advancements in the use of frozen–thawed semen for honey bee insemination, particularly noting the potential for successful female brood production without the addition of antibiotics and the evaluation of viability by different combinations of dyes. While the observed outcomes demonstrate comparable queen mortality rates between frozen and fresh insemination techniques, the implications of low queen weight and limited egg-laying periods warrant further investigation. The findings emphasize the importance of refining cryopreservation protocols, including optimal cryoprotectant concentrations and minimizing contamination during semen collection, to enhance the viability and longevity of the sperm and also of the inseminated queens.

Author Contributions: Conceptualization, S.E. and S.D.; methodology, S.E. and S.D.; investigation, S.E. and B.T.; resources, F.E., J.P. and B.T.; writing—original draft preparation, S.E.; writing—review and editing, S.E., S.D., F.E., J.P. and B.T.; supervision, S.D.; project administration, S.E.; funding acquisition, S.E. and S.D. All authors have read and agreed to the published version of the manuscript.

Funding: This research was partly funded by Fonds National de la Recherche Scientifique—FNRS, grant number 1V20123F VETE-CCD, by a provision of a PhD Studentship and partly by the Faculty of Veterinary Medicine from the Universtity of Liège through a grant awarded by les Fonds Spéciaux de la Recherche—FSR-F-2021-FreezeBee.

Data Availability Statement: Data are contained within the article.

Acknowledgments: The authors thank Alexandre Hego and Gaëtan Lefevre from the epifluorescence and flow cytometry platform of the GIGA of the University of Liège for their technical support and Rene Van Der Molen for drone semen collection. The graphical abstract was Created in BioRender. https://BioRender.com/i71j767.

Conflicts of Interest: The authors declare no conflicts of interest.

References

1. Aizen, M.A.; Harder, L.D. The Global Stock of Domesticated Honey Bees Is Growing Slower Than Agricultural Demand for Pollination. *Curr. Biol.* **2009**, *19*, 915–918. [CrossRef]
2. Potts, S.G.; Biesmeijer, J.C.; Kremen, C.; Neumann, P.; Schweiger, O.; Kunin, W.E. Global Pollinator Declines: Trends, Impacts and Drivers. *Trends Ecol. Evol.* **2010**, *25*, 345–353. [CrossRef]
3. Weybright, S. Importing Frozen Honey Bee Sperm Is Key to Conservation. Available online: https://news.wsu.edu/2016/07/25/importing-frozen-honey-bee-sperm-key-conservation/ (accessed on 27 January 2019).
4. Metcalf, E.S. The Role of International Transport of Equine Semen on Disease Transmission. *Anim. Reprod. Sci.* **2001**, *68*, 229–237. [CrossRef]
5. Yue, C.; Schröder, M.; Bienefeld, K.; Genersch, E. Detection of Viral Sequences in Semen of Honeybees (*Apis mellifera*): Evidence for Vertical Transmission of Viruses through Drones. *J. Invertebr. Pathol.* **2006**, *92*, 105–108. [CrossRef]
6. Wegener, J.; May, T.; Knollmann, U.; Kamp, G.; Müller, K.; Bienefeld, K. In Vivo Validation of in Vitro Quality Tests for Cryopreserved Honey Bee Semen. *Cryobiology* **2012**, *65*, 126–131. [CrossRef]
7. Gül, A.; Şahinler, N.; Onal, A.G.; Hopkins, B.K.; Sheppard, W.S. Effects of Diluents and Plasma on Honey Bee (*Apis mellifera* L.) Drone Frozen-Thawed Semen Fertility. *Theriogenology* **2017**, *101*, 109–113. [CrossRef]
8. Hopkins, B.K.; Herr, C.; Sheppard, W.S. Sequential Generations of Honey Bee (*Apis mellifera*) Queens Produced Using Cryopreserved Semen. *Reprod. Fertil. Dev.* **2012**, *24*, 1079–1083. [CrossRef]
9. Wegener, J.; May, T.; Kamp, G.; Bienefeld, K. A Successful New Approach to Honeybee Semen Cryopreservation. *Cryobiology* **2014**, *69*, 236–242. [CrossRef]
10. Auth, C.A.; Hopkins, B.K. Nitrogen Vapor Immersion: An Accessible Alternative for Honey Bee (*Apis mellifera* L.) Semen Cryopreservation. *Cryobiology* **2021**, *100*, 12–18. [CrossRef]
11. Taylor, M.A.; Guzmán-Novoa, E.; Morfin, N.; Buhr, M.M. Improving Viability of Cryopreserved Honey Bee (*Apis mellifera* L.) Sperm with Selected Diluents, Cryoprotectants, and Semen Dilution Ratios. *Theriogenology* **2009**, *72*, 149–159. [CrossRef]
12. Paillard, M.; Rousseau, A.; Giovenazzo, P.; Bailey, J.L. Preservation of Domesticated Honey Bee (Hymenoptera: Apidae) Drone Semen. *J. Econ. Entomol.* **2017**, *110*, 1412–1418. [CrossRef]

13. Kaftanoglu, O.; Peng, Y.S. Preservation of Honeybee Spermatozoa in Liquid Nitrogen. *J. Apic. Res.* **1984**, *23*, 157–163. [CrossRef]
14. Nur, Z.; Çakmak, S.S.; Çakmak, I.; Önder, N.T.; Gökçe, E.; Üstüner, B.; Alçay, S.; Toker, M.B.; Şen, H.; Soylu, M.K. Effect of Partial Replacement of Dimethyl Sulfoxide with Dextran, Polyvinylpyrrolidone or Polyethylene Glycol on Post-Thaw Drone Semen Quality and Queen Fertility. *J. Hell. Vet. Med. Soc.* **2024**, *75*, 7371–7378. [CrossRef]
15. Hopkins, B.K.; Herr, C. Factors Affecting the Successful Cryopreservation of Honey Bee (*Apis mellifera*) Spermatozoa. *Apidologie* **2010**, *41*, 548–556. [CrossRef]
16. Alcay, S.; Cakmak, S.; Cakmak, I.; Mulkpinar, E.; Gokce, E.; Ustuner, B.; Sen, H.; Nur, Z. Successful Cryopreservation of Honey Bee Drone Spermatozoa with Royal Jelly Supplemented Extenders. *Cryobiology* **2019**, *87*, 28–31. [CrossRef]
17. Rajamohan, A.; Danka, R.G.; Hopkins, B.K.; Rinehart, J.P. A Non-Activating Diluent to Prolong in Vitro Viability of *Apis mellifera* Spermatozoa: Effects on Cryopreservation and on Egg Fertilization. *Cryobiology* **2020**, *92*, 124–129. [CrossRef]
18. Gulov, A.N.; Berezin, A.S.; Larkina, E.O.; Saltykova, E.S.; Kaskinova, M.D. Creation of a Biobank of the Sperm of the Honey Bee Drones of Different Subspecies of *Apis mellifera* L. *Animals* **2023**, *13*, 3684. [CrossRef]
19. Morrell, J.M.; Cojkic, A.; Malaluang, P.; Ntallaris, T.; Lindahl, J.; Hansson, I. Antibiotics in Semen Extenders—A Multiplicity of Paradoxes. *Reprod. Fertil. Dev.* **2024**, *36*, RD23218. [CrossRef]
20. World Health Organization. *WHO Laboratory Manual for the Examination and Processing of Human Semen Sixth Edition*; Björndahl, L., Apolkhin, O., Baldi, E., Barratt, C., Festin, M., Kirkman-Brown, J., Lamb, D., Mbizvo, M., Schlatt, S., Toskin, I., et al., Eds.; World Health Organization: Geneva, Switzerland, 2021; Volume V, p. 286, ISBN 978 92 4 154778 9.
21. Smilga-Spalvina, A.; Spalvins, K.; Veidenbergs, I. Review of Sustainable Cryopreservation and Above-Freezing Storage Solutions of European Honey Bee *Apis mellifera* Drone Semen. *Environ. Clim. Technol.* **2023**, *27*, 177–194. [CrossRef]
22. Egyptien, S.; Dewals, B.; Ectors, F.; Brutinel, F.; Ponthier, J.; Deleuze, S. Validation of Calcein Violet as a New Marker of Semen Membrane Integrity in Domestic Animals. *Animals* **2023**, *13*, 1874. [CrossRef]
23. Wegener, J.; Bienefeld, K. Toxicity of Cryoprotectants to Honey Bee Semen and Queens. *Theriogenology* **2012**, *77*, 600–607. [CrossRef]
24. Cobey, S.W.; Tarpy, D.R.; Woyke, J. Standard Methods for Instrumental Insemination of *Apis mellifera* Queens. *J. Apic. Res.* **2013**, *52*, 1–18. [CrossRef]
25. Pinto, M.A.; Yániz, J.; Toquet, M.; Santolaria, P.; Silvestre, M.A.; Toledo-Perona, R.; Gómez-Martín, Á. Microbiota Analysis of Ejaculated Honey Bee Drone Semen and the Effect of Semen Collection Method on Bacterial Loads. *Insects* **2024**, *15*, 377. [CrossRef]
26. Farahani, L.; Tharakan, T.; Yap, T.; Ramsay, J.W.; Jayasena, C.N.; Minhas, S. The Semen Microbiome and Its Impact on Sperm Function and male Fertility: A Systematic Review and Meta-Analysis. *Andrology* **2021**, *9*, 115–144. [CrossRef]
27. Akyol, E.; Yeninar, H.; Kaftanoglu, O. Live Weight of Queen Honey Bees (*Apis mellifera* L.) Predicts Reproductive Characteristics. *J. Kans. Entomol. Soc.* **2008**, *81*, 92–100. [CrossRef]
28. Collins, A.M.; Pettis, J.S. Correlation of Queen Size and Spermathecal Contents and Effects of Miticide Exposure during Development. *Apidologie* **2013**, *44*, 351–356. [CrossRef]
29. Kahya, Y.; Gençer, H.V.; Woyke, J. Weight at Emergence of Honey Bee (*Apis mellifera* Caucasica) Queens and Its Effect on Live Weights at the Pre and Post Mating Periods. *J. Apic. Res.* **2008**, *47*, 118–125. [CrossRef]
30. Vasfi Gençer, H.; Shah, S.Q.; Firatli, Ç. Effects of Supplemental Feeding of Queen Rearing Colonies and Larval Age on the Acceptance of Grafted Larvae and Queen Traits. *Pak. J. Biol. Sci.* **2000**, *3*, 1319–1322. [CrossRef]
31. Büchler, R.; Andonov, S.; Bernstein, R.; Bienefeld, K.; Costa, C.; Du, M.; Gabel, M.; Given, K.; Hatjina, F.; Harpur, B.A.; et al. Standard Methods for Rearing and Selection of *Apis mellifera* Queens 2.0. *J. Apic. Res.* **2024**. [CrossRef]
32. Collins, A. Relationship between Semen Quality and performance of Instrumentally Inseminated Honey Bee Queens. *Apidologie* **2000**, *31*, 421–429. [CrossRef]
33. Yániz, J.; Palacín, I.; Santolaria, P. Effect of Chamber Characteristics, Incubation, and Diluent on Motility of Honey Bee (*Apis mellifera*) Drone Sperm. *Apidologie* **2019**, *50*, 472–481. [CrossRef]
34. Divasón, J.; Romero, A.; Silvestre, M.A.; Santolaria, P.; Yániz, J.L. In Vitro Maintenance of Drones and Development of a New Software for Sperm Quality Analysis Facilitate the Study of Honey Bee Reproductive Quality. *J. Apic. Res.* **2023**, *63*, 1088–1095. [CrossRef]

insects

MDPI

Review

An Overview of the Adverse Impacts of Old Combs on Honeybee Colonies and Recommended Beekeeping Management Strategies

Qingxin Meng †, Rong Huang †, Shunhua Yang, Wutao Jiang, Yakai Tian * and Kun Dong *

Yunnan Provincial Engineering and Research Center for Sustainable Utilization of Honeybee Resources, Eastern Bee Research Institute, College of Animal Science and Technology, Yunnan Agricultural University, Kunming 650201, China

* Correspondence: tykzhong@foxmail.com (Y.T.); dongkun19722004@aliyun.com (K.D.)

† These authors contributed equally to this work.

Simple Summary: Honeycombs, where bees raise their offspring and store food, naturally age over time. Old combs become darker, accumulate harmful substances like pesticides and heavy metals, and undergo structural changes that result in the shrinkage of the cells where bees develop. These changes lead to smaller workers with shorter lifespans and reduced abilities to forage, thereby weakening the entire colony's performance. Additionally, the physical and chemical properties of honey and beeswax change as the honeycombs age. To mitigate the negative impacts of old combs, bees engage in behaviors to clean or rebuild combs, and at the body level, systemic molecular mechanisms of immune system adaptation and detoxification take place. This review highlights the importance of regularly replacing old brood combs in beekeeping practices to maintain healthy colonies, improve honey production, and reduce environmental pollution risks. By managing comb age, beekeepers can support stronger colonies and contribute to sustainable agriculture and ecosystem conservation.

Abstract: The honeybee comb serves as the primary site for all essential colony activities, and its structural and functional integrity plays a crucial role in colony development. As combs age through successive brood-rearing cycles, their physicochemical properties undergo significant changes, which can negatively affect colony health and productivity. This review synthesizes the current knowledge on the biological functions of combs, the aging process, and the negative impacts of old combs on cell structure, worker morphology, colony strength, and bee product quality. Additionally, it examines the adaptive strategies employed by honeybees and the recommended beekeeping management practices to mitigate these effects. Specifically, old combs undergo structural changes in cell dimensions and reduced spatial capacity, leading to the growth of smaller bees with diminished foraging and productivity. Furthermore, bee products, such as honey and beeswax, collected from old combs demonstrate compromised quality and higher levels of environmental contaminants. To counteract these challenges, colonies engage in hygienic behaviors, such as cell cleaning and comb rebuilding, while their enhanced immune and detoxification systems help mitigate comb-derived stressors. This review demonstrates that the systematic replacement of old brood combs is a critical management strategy to optimize bee health and ensure sustainable apiculture.

Keywords: old comb; worker development; colony strength; immune detoxification; beekeeping management

https://doi.org/10.3390/insects16040351

1. Introduction

Honeybee colonies represent a cornerstone of global ecosystems and agricultural productivity, serving as indispensable pollinators for both wild flora and cultivated crops [1]. At the heart of colony functionality lies the comb—a dynamic structure meticulously engineered by bees to support brood rearing, food storage, and social communication [2]. Constructed from beeswax secreted by workers, combs undergo progressive physicochemical alterations through repeated brood-rearing cycles or from stored honey and pollen [3,4], yet the implications of these aging processes on colony health remain underexplored. While extensive research has elucidated the comb's role in thermoregulation and nutrient storage [5,6], critical gaps persist in understanding how structural degradation and pollutant accumulation in old combs compromise individual bee development, colony resilience, and apicultural output.

Recent studies highlight alarming trends: global bee populations face unprecedented declines due to anthropogenic stressors, including habitat loss, pesticide exposure, and climate change [7,8]. Within this context, the comb emerges not only as a habitat but also as a reservoir for environmental contaminants, with old combs accumulating heavy metals, pesticides, and microbial pathogens [9,10]. These contaminants synergize with structural decay—such as reduced cell volume and thickened walls—to constrain larval development, diminish worker size, and impair foraging efficiency [11,12]. Furthermore, degraded combs compromise the quality of hive products, raising concerns for food safety and economic viability [13,14]. Despite these challenges, bees exhibit remarkable adaptive behaviors, including comb rebuilding and detoxification mechanisms, which remain poorly integrated into current apicultural practices [8,15].

This review synthesizes multidisciplinary advances to address three pivotal questions: (1) How do old combs structurally and chemically change, and what are the cascading effects on bee ontogeny and colony dynamics? (2) What adaptive strategies do bees employ to mitigate comb-derived stressors, and how can these inform sustainable management? (3) What innovations in comb renewal protocols and pollutant monitoring are critical for safeguarding bee health? By integrating empirical and experimental data from ecological and apicultural studies, this work elucidates the interplay between comb aging, colony fitness, and environmental pressures. It further proposes evidence-based strategies for comb management, emphasizing timely replacement and pollutant mitigation to enhance apicultural resilience. As pollinator declines threaten global food security, this synthesis underscores the urgency of prioritizing comb health as a linchpin for sustainable beekeeping and ecosystem conservation.

2. Biological Functions of Combs

The honeybee comb is a multifunctional biosynthetic architecture central to colony survival, serving as the structural and physiological cornerstone for brood development and social coordination [2]. Combs are constructed from beeswax, a complex lipids and hydrocarbons mixture secreted by the wax glands of workers. These glands, located on the ventral abdomen, convert carbohydrates from honey into wax scales through a series of enzymatic processes [16]. A honeybee nest (*Apis* spp.) is composed of one or more vertically hanging combs, spaced apart and arranged in parallel, which are crucial for maintaining the stability of the internal environment of the colony [17]. Within the dark confines of the hive, bees use the "dance language" to communicate with their peers on the comb surface about the distance and location of food resources [18].

The fundamental structural unit of the comb is the cell, which is classified into queen, worker, and drone cells based on size and shape [2]. Queen cells, specialized for queen rearing, are cup-shaped structures typically located along the lower edges of the comb, with their openings facing downward and perpendicular to the ground [19]. Worker and drone cells, primarily used for rearing workers and drones, respectively, are hexagonal prisms with bases composed of three rhombic plates [20]. The majority of the cells in the comb are worker cells, whose primary function is to provide a stable and suitable environment for the growth and development of workers. Additionally, these cells located at the upper edges of the comb can also be used to store honey and pollen, serving as a nutrient supplement for bees [21].

Furthermore, the comb and its byproducts hold significant value for human use. Beeswax is rich in bioactive compounds with therapeutic properties, including anti-inflammatory effects beneficial for conditions, such as rhinitis and hepatitis [22]. Additionally, it exhibits antioxidant, antibacterial, analgesic, and immune-enhancing activities [23]. Beyond its medicinal applications, beeswax is used as a raw material in different industries, such as coatings for food and pharmaceuticals, and cosmetics, etc. [24].

3. Aging Process of the Comb

The aging of the comb is a gradual process, involving changes in both physical properties and biological functions. One of the most apparent phenotypic changes in old combs is the darkening of their color, which appears as deep brown or black (Figure 1A) [25]. Additionally, old combs influence cell structure, the morphological size of the workers reared within them, and overall colony productivity (Figure 1B,C) [3]. Old combs also tend to accumulate various harmful substances, including fungi, bacteria, pesticides, and heavy metals, which can negatively affect individual bee health and colony development [9,10].

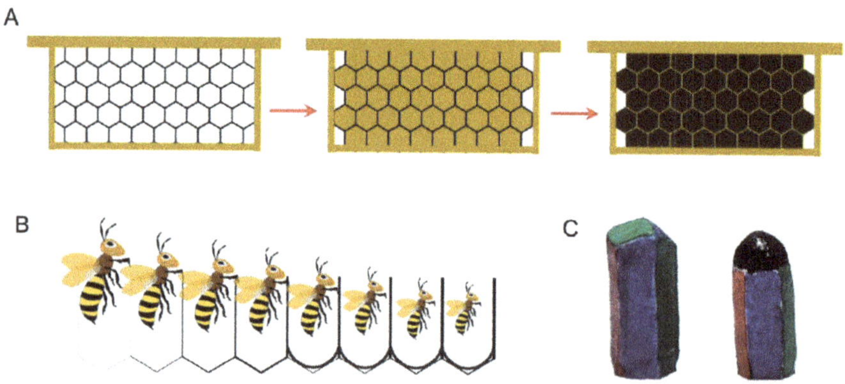

Figure 1. With the increase in the number of brood-rearing generations, (**A**) the comb color gradually darkens to black, (**B**) the cell volume decreases and the size of the reared worker reduces, and (**C**) the cell structure undergoes changes.

3.1. Color Changes in the Comb

Beeswax, the primary component of the comb, readily adsorbs pigments from accumulated residues within the cells. Its porous structure provides numerous adsorption sites, allowing pigment particles to penetrate and adhere to the surface or interior of the wax [26]. In addition, the main constituents of beeswax include fatty acid esters (~67%), hydrocarbons (~14%), and free fatty acids (~13%) [24]. These compounds are rich in polar groups that can interact with pigments through hydrogen bonding or van der Waals forces, resulting in strong adsorption [27,28]. As combs store nectar and pollen, they gradually darken in color from their initial white or light yellow to deeper shades. This color change is also influenced by physiological activities occurring within the cells, such as fecal excretion, silk spinning for cocoon formation, and molting during bee development [29]. In particular, bee feces contain dark pigments that are easily absorbed by beeswax [11,30]. With successive brood-rearing generations, the accumulation of cocoons and other debris causes old combs to turn brown or black [25].

3.2. Structural Changes in the Comb Cells

Cells are the fundamental units of the comb, serving as primary sites for food storage, brood development, and adult bee rest, all of which are critical for colony development [31]. Newly constructed comb cells are typically hexagonal prisms with a base formed by three rhombic plates [32]. This design provides structural advantages, including lightweight construction, high load-bearing capacity, and space efficiency [2,17]. For instance, 1 g of beeswax can produce approximately 20 cm^2 of double-sided comb surface, while only 55 g of wax is required to build a comb capable of supporting 1 kg of ripened and capped honey [33]. However, brood rearing leads to structural changes in the cells due to the accumulation of metabolic byproducts. With each successive generation of workers reared in a cell, residues, such as feces and cocoon silk, accumulate as they cannot be fully removed by nurse bees [11]. This accumulation increases comb weight and thickens the cell walls. Notably, the dimensions of comb cell structures vary among different bee species, leading to species-specific fluctuations in the observed changes (Table 1). For instance, the weight of a newly constructed comb of *A. m. carnica* (housed in Kafr El Sheikh, Egypt) is 0.26 g/cm^2, but as the comb ages from 1 to 7 years, its weight gradually increases to 0.33, 0.60, 0.88, 0.95, 1.04, 1.17, and 1.32 g/cm^2, respectively [34]. Similarly, the thickness of newly constructed cell walls in the upper part is 88 μm, but after 5 months, 1 year, and 2 years of brood rearing, it increases to 120 μm, 246 μm, and 297 μm, respectively [12].

Table 1. Diameter of brood cells of each bee species in the genus *Apis*.

Honeybee Species		Average Cell Diameter (mm)		Citation
		Worker	Drone	
Apis mellifera	*A. m. carnica*	5.51	6.91	Ruttner, 1988 [35]
	A. m. ligustica	5.49	6.62	Yang et al., 2021 [20]
	A. m. scutellata	4.80	6.15	Smith et al., 1961 [36]
	A. m. jemenitica	4.75	6.19	Dutton et al., 1981 [37]
	A. m. litorea	4.62	6.15	Ruttner, 1988 [35]
Apis cerana	*A. c. cerana*	4.82	5.62	Yang et al., 2021 [20]
	A. c. japonica	4.78	/	Ruttner, 1988 [35]
	A. c. indica	4.37	/	Ruttner, 1988 [35]
Apis dorsata		5.32	/	Khan et al., 2007 [38]
Apis laboriosa		6.3	/	Starr, 2021 [39]
Apis florea		2.7–3.1	4.2–4.8	Free, 1981 [40]
Apis andreniformis		2.8	4.2	Starr, 2021 [39]

The increase in comb weight and cell wall thickness also alters the internal shape and structural dimensions of the cells. Over time, the hexagonal prism shape of the cells transforms into a cylindrical form with a hemispherical base [32]. Compared with newly constructed brood cells, old brood cells have reduced diameter and volume [34]. For *A. m. carnica* (housed in Kafr El Sheikh, Egypt), the diameter and volume of newly constructed worker cells were measured as 6.00 mm and 0.31 mL, respectively. After 6 years of use, these dimensions decreased to 4.86 mm and 0.18 mL. A similar trend was observed in *A. c. cerana* [25]. The primary factor driving these changes is the accumulation of cocoon residue within the cells. For example, in old brood cells of *A. c. cerana* (housed in Kunming, China), the weight of the cocoon is 57.90 mg, and the thickness at the base is 0.96 mm, occupying approximately 13% of the cell space and significantly reducing the base dimensions [11]. Therefore, as the number of worker generations reared in the comb increases, the accumulation of the cocoon increases the comb weight and reduces the cell dimensions, ultimately restricting the developmental space available for bees within the cells.

3.3. *Effects of Old Combs on the External Morphological Traits of the Workers*

The structural dimensions of the brood cells are significantly correlated with the morphological characteristics of honeybees [41,42]. Old brood cells are smaller in volume compared with newly constructed cells, resulting in reduced nutrient provisioning [43,44]. Within the fixed developmental period, bee larvae reared in old cells are unable to acquire sufficient nutrients for full maturation [45]. The smaller dimensions of the old cells impose spatial stress during worker development [11]. For instance, *A. m. carnica* (housed in Somogy, Hungary) pupae that were reared in new cells exhibited significantly larger head (1.41 vs. 1.16 mm), thorax (4.74 vs. 3.60 mm), and abdomen dimensions (5.03 vs. 4.15 mm) and higher birth weights (114.89 vs. 88.05 mg) compared with those reared in old brood cells [4,46]. The birth weight is an important indicator of bee development, reflecting the size of the bee's external morphological structure. The external morphological dimensions of bees are negatively correlated with the age of the comb in which they develop [25]. In *A. m. carnica* (housed in Hofuf, Saudi Arabia), the proboscis length of the workers reared in combs aged 1–4 years was 6.10, 5.94, 5.55, and 5.30 mm, respectively; the forewing length was 9.05, 9.03, 9.00, and 9.00 mm; the length of the third tergite was 2.30, 2.25, 2.10, and 2.05 mm; the tarsus length was 2.50, 2.48, 2.45, and 2.45 mm; and the wax mirror length was 1.55, 1.55, 1.51, and 1.50 mm [47].

The structural dimensions of the cells can influence the external morphological dimensions of bees, and vice versa. Li reported that small-sized bees reared in old combs constructed new cells with significantly reduced diameters and depths compared with large-sized bees reared in new combs [42]. Within the colony, each wax-producing bee contributes to beeswax secretion and processing, using its antennae to identify the location and quantity of wax deposition [2]. During cell construction, bees typically use their legs and antennae as measuring tools. Consequently, small-sized workers construct smaller cells due to their reduced morphological structures [48]. Additionally, the external morphology of bees is closely related to their foraging and production capabilities. Thus, the reduction in individual bee size observed in colonies using old combs may directly impair both individual foraging capacity and overall colony productivity [4,49]. However, further experimental validation is required to confirm these causal relationships.

3.4. *Impact of Old Combs on Colony Strength and Production Performance*

Old combs negatively impact colony strength and production performance. Colony strength, defined as the number of workers within a colony, is a key indicator of reproductive and productive capacity [50]. Old comb significantly reduces the number of newly emerged workers and shortens their lifespan. In *A. m. carnica* (housed in Damietta, Egypt), the area of sealed broods in combs aged 1–4 years was 3722, 3698, 2985, and 2413 cm^2, respectively; the survival rate of the sealed brood was 74.45, 71.05, 72.60, and 72.50%; and the average lifespan of the newly emerged workers was 30.0, 29.7, 27.4, and 24.0 d [3]. Therefore, colonies with predominantly new combs exhibit significantly better colony strength than those with old combs. A robust colony ensures better internal organization, with each bee performing its role effectively, leading to improved production performance [3]. The foraging frequency of workers is one of the primary factors influencing colony production performance. For example, *A. m. carnica* (housed in Al-Ahsa oasis, Saudi Arabia) colonies with 1-year-old combs had an average of 19.85 workers returning with pollen per minute, which was significantly higher than that of colonies with 2-year-old combs, which was 16.95 [51]. This may be due to the larger size and better developmental condition of workers reared in new combs, resulting in differences in flight speed and foraging efficiency. Similarly, colonies established with new or old combs exhibit differences in food storage capacity. *Apis m. carnica* (housed in Kafr El Sheikh, Egypt) colonies consisting of new combs stored 159.89 and 266.55 cm^2 of pollen and honey, respectively, whereas those with old combs stored 95.55 and 141.16 cm^2, respectively (Figure 2) [4].

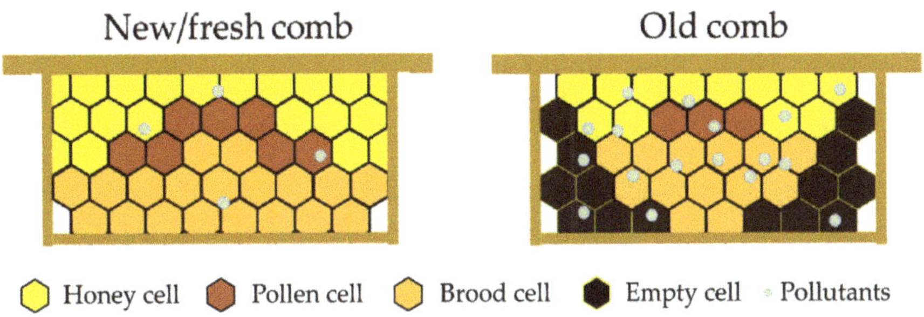

Figure 2. Compared with new/fresh combs, old combs exhibit a decrease in brood, nectar, and pollen, while showing an increased accumulation of pollutants.

3.5. Impact of Old Combs on the Quality of Bee Products

The honey quality changes with the age of the comb. A study conducted at Kafr El-Sheikh University on *A. m. carnica* revealed that honey stored in 4-year-old combs exhibited increased specific gravity and viscosity compared to fresher combs [14]. Furthermore, the chemical composition of honey shifted with comb age, with reducing sugar content rising and glucose levels declining. Organoleptic evaluation, a direct method for measuring, analyzing, and interpreting food quality, revealed that honey stored in old combs is less acceptable in terms of color, odor, taste, viscosity, and sourness. The hydroxymethylfurfural (HMF) content is an important indicator of honey quality and is used to assess its freshness [52]. The HMF content in honey produced in old combs (28.8 mg/kg) was found to be significantly higher than that in honey produced in new combs (18.4 mg/kg), potentially affecting honey quality [53]. Additionally, combs can be a reservoir for various metal elements, and concentrations of K, Na, Ca, Mg, Fe, Zn, Pb, and Mn in honey collected from different combs significantly increase with comb age [54]. Furthermore, beeswax undergoes changes in quality with comb age. Old combs have fewer long-chain esters and more short-chain compounds, with the content of alkenes, dienes, and branched alkanes being significantly higher than in new combs, while that of alkanes decreases. Due to the higher volatility of alkanes compared with alkenes, alkanes gradually volatilize as the comb age increases, representing a characteristic change in the beeswax over time [55].

3.6. Accumulation of Environmental Pollutants in Old Combs

Bees play a vital role in terrestrial ecosystems as primary pollinators for most crops and wild plants, making them essential for maintaining wild plant communities and agricultural productivity [56]. Bee pollination significantly enhances the fruit set rate, seed set rate, yield, and quality of blueberries [57]. In 2009, bee pollination contributed an estimated $11.68 billion to agriculture in the United States [1]. However, bee populations have been declining due to global climate change, habitat fragmentation, and harmful substances released from industrial and agricultural activities, such as pesticides [8]. Foraging bees may collect food contaminated with heavy metals, pesticides, and other harmful substances, while airborne particulate pollutants can adhere to their body hairs [58]. These contaminants readily integrate with beeswax, accumulating over time and increasing in concentration as the comb ages [59]. Consequently, old combs accumulate higher levels of environmental pollutants (Figure 2). The concentrations of Cd, Cr, Ni, Pb, and Mn in combs aged 5 years increased by 56%, 63%, 65%, 78%, and 82%, respectively, compared with 1-year-old combs [9]. Additionally, the cocoons in old combs maintain close contact with each worker for at least 20 d during development [29,30]. Therefore, the concentration of heavy metals in these cocoons also reflects the negative impact of old combs on bee health [60]. Specifically, the concentrations of Cr, Cd, Pb, Mn, Ni, and As in cocoons derived from old combs of *A. c. cerana* were significantly higher than those in newly built brood cells. Moreover, the heavy metal content and detoxification gene expression in 6-day-old larvae and newly emerged workers developing in these old combs were significantly increased.

The hydrophobic-lipophilic properties of comb constituents create differential environmental pollutant accumulation patterns between hive products. Due to their hydrophilic characteristics, certain contaminants preferentially partition into honey matrices [61]. Conversely, the lipophilic nature of beeswax and pollen facilitates the bioaccumulation of persistent organic pollutants, including miticides and veterinary pharmaceuticals used in apiculture [62]. This makes honeycomb a reliable bioindicator matrix for environmental monitoring, enabling simultaneous assessment of ecosystem contamination levels and apian population exposure risks [63]. For example, a study monitoring 11 heavy metals in bee products from colonies in the Campania region of Italy across three time periods (T1: COVID-19 pandemic lockdown; T2: restricted activity; T3: full reopening) found that the heavy metal content in bee products during the lockdown was significantly lower than that during the restricted activity and full reopening periods. This indicates that limiting industrial activities can help reduce pollutant levels both in the environmental and within bee colonies [64]. Similarly, the levels of environmental pollutants in colonies vary with human activity density, with higher Hg and Pb content found in honey from urban areas compared with that in mountainous regions [65]. Pollutants are ubiquitous in the environment, and when environmental pollutant levels are high, they are transferred into the hive, leading to increased pollutant levels in the combs, bee products, and bees themselves, which severely negatively impacts the quality of bee products and colony health [66,67]. Moreover, a critical concern is the widespread practice in the beekeeping industry of recycling beeswax from old combs into wax foundation sheets, which are reintroduced into hives [68]. These sheets often contain higher pollutant loads (like acaricides), and when provided to honeybee colonies, they may compromise colony health and survivorship [69].

4. Bee Adaptations in Old Combs

4.1. Behavioral Adaptations to Structural Degradation

Old combs impose significant negative impacts on individual bees, colony strength, and production performance. Thus, bee colonies may respond with different behaviors to mitigate the threats posed by old combs. For instance, in *A. mellifera* (housed in Moscow, Russia), the volume of brood cells decreases with successive generations of brood rearing. However, when the cell volume shrinks to 88% of that of new cells, workers meticulously clean the deposited residues within the cells and secrete wax to heighten the cell walls to reduce any further reduction and ensure minimum spatial requirements for normal brood development [70]. This cleaning and cell-wall extension strategy allows *A. mellifera* to counteract the developmental constraints caused by cell volume reduction in old combs. The Eastern honeybee (*A. cerana*) has evolved a distinct strategy, gnawing on old combs (Figure 3) [15]. Gnawing marks are frequently observed on the surface of old combs in *A. c. cerana* colonies, with abundant wax debris accumulating at the hive bottom. During periods of abundant nectar and pollen resources, workers rebuild the gnawed areas by secreting fresh wax, forming entirely new or semi-reconstructed cells. The dimensions of these semi-reconstructed cells and the morphological traits of the workers reared within them are significantly larger than those in old brood cells, demonstrating that gnawing and rebuilding old brood cells is an effective comb renewal strategy [11].

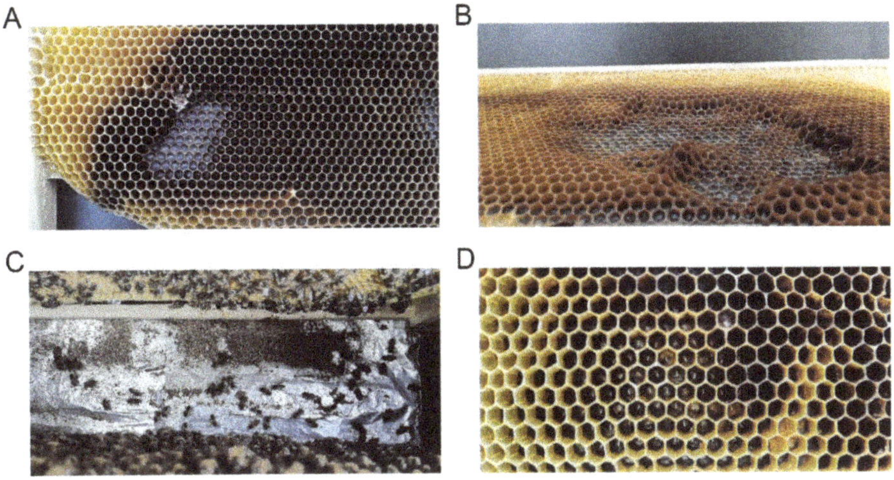

Figure 3. (**A**,**B**) Gnawed regions on old combs of *A. c. cerana*. (**C**) Wax residues accumulation at the hive bottom. (**D**) Comb cell reconstruction via wax secretion in gnawed areas.

4.2. Molecular Mechanisms of Pollutant Detoxification

To combat environmental pollutants, bees have evolved detoxification enzyme gene families that metabolize both endogenous compounds and xenobiotics, including heavy metals, pesticides, and chemical carcinogens [71]. The primary detoxification enzyme superfamilies in bees include cytochrome P450 monooxygenases, carboxylesterases, and glutathione S-transferases [8]. Among these, the CYP450 gene family plays a crucial role in enhancing insect tolerance to environmental stressors and is recognized as a biomarker for pollutant exposure [65]. CYP450 monooxygenases exhibit peroxidase-like activity, protecting against reactive oxygen species (ROS) [72]. Glutathione S-transferases conjugate glutathione's thiol group to ROS or electrophilic compounds, increasing their hydrophilicity and promoting cellular elimination [73]. The expression of these enzymes is tightly regulated by transcription factors and signaling pathways, including AhR/ARNT, CncC/Keap1, GPCR, and MAPK/CREB, which enhance detoxification capacity and stress tolerance [74].

5. Beekeeping Management Practices

Old combs negatively impact colony health and productivity. Therefore, to safeguard bee populations and promote efficient apiculture, targeted management practices are essential. Researchers recommend timely replacement of old combs to mitigate their negative impacts. In *A. mellifera*, removing combs older than 3 years facilitates the rearing of larger workers and supports optimal productivity [3,4]. In *A. cerana*, combs older than 6 months or those used for over eight brood generations should be replaced to promote individual development and colony growth [25]. Delayed comb replacement in *A. cerana* colonies can trigger comb-gnawing behavior, leading to wax debris accumulation. This debris creates an ideal environment for wax moths (*Galleria mellonella*) to lay eggs and fosters larval infestation [75]. Wax moth larvae burrow into old combs, excavating tunnels that result in pupal mortality [76]. Therefore, infestations can impair colony reproduction and may lead to absconding, causing substantial economic losses. Thus, beekeepers are advised to implement regular comb replacement protocols. When replacement is delayed, thorough hive cleaning is imperative to eliminate pest infestations and harmful microbial infections.

Importantly, the replacement of old combs necessitates proper disposal to mitigate potential contamination risks. These aged combs are typically processed for beeswax recovery, which is subsequently reintroduced into industries utilizing wax-based products, such as the production of foundation sheets to support colony development [68]. To ensure the safety and quality of recycled beeswax, appropriate physicochemical treatments are essential to reduce or eliminate contaminant levels. Recent studies have demonstrated that beeswax filtered using Norit SA4 PAH-activated carbon and Tonsil 114 FF diatomaceous earth achieves significant reductions in acaricides (e.g., chlorfenvinphos, coumaphos) and heavy metals (e.g., arsenic, mercury), while the hydrocarbon and monoester contents remain statistically unaltered [77]. However, it is noteworthy that total phenolic compounds, flavonoids, and the antioxidant capacity of lipophilic extracts exhibit slight reductions following treatment. Furthermore, an extraction process employing methanol at 65 °C, followed by an adequation process with water at 70 °C, has been shown to effectively reduce contamination levels of acaricides and pesticides in beeswax [78]. The decontaminated beeswax produced through this method exhibits quality parameters comparable to those of virgin beeswax. Notably, 80% of the methanol used in the extraction process is recyclable, and the resulting waste is manageable with minimal environmental impact. Despite these advancements, the development of advanced decontamination methodologies with superior comprehensive efficacy, minimized residual contamination, and enhanced processing efficiency remains a critical research priority. Further in-depth investigations are warranted to address these challenges and optimize the sustainability of beeswax recycling processes.

6. Conclusions

As the foundation of bee life, combs directly influence colony reproduction and productivity. Prolonged comb usage alters their physical and chemical properties, affecting individual development and overall colony health. Structural degradation of old combs results in changes to worker morphology and reduces foraging efficiency. Furthermore, pollutant accumulation in old combs negatively impacts bee health and compromises product quality. Despite these challenges, bees exhibit adaptive strategies. *Apis mellifera* maintains cell functionality through cleaning and wall extension, while *A. cerana* rebuild cells by gnawing on old combs. These behaviors underscore the resilience of honeybees and can inform apicultural management practices. Regular comb replacement reduces pollutant accumulation, improves developmental conditions, and enhances colony performance. With the decline in bee populations, identifying threats remains imperative. Future research should explore the application of advanced detection technologies, molecular biology, and multi-omics approaches to better understand pollutant profiles, their impacts on bee health, and bee resistance mechanisms. Such insights could guide strategies for improving bee health, such as identifying biomarkers or modulating gut microbiota. Furthermore, it is worth further exploring the efficient decontamination method of beeswax. In conclusion, maintaining high-quality combs is essential for colony vitality. Therefore, effective comb management not only boosts apicultural efficiency but also contributes to biodiversity conservation and ecological balance.

Author Contributions: Conceptualization, K.D.; methodology, Y.T.; writing—original draft preparation, Q.M. and R.H.; writing—review and editing, Y.T. and K.D.; visualization, S.Y. and W.J.; project administration, K.D.; funding acquisition, K.D. All authors have read and agreed to the published version of the manuscript.

Funding: This research was funded by National Natural Science Foundation of China, grant numbers 32060241 and 31572339; China Agriculture Research System of MOF and MARA, grant number CARS-44-KXJ13; Reserve Talents Training Program for Young and Middle-aged Academic and Technical Leaders in Yunnan, grant number 2018HB041.

Institutional Review Board Statement: Not applicable.

Informed Consent Statement: Not applicable.

Data Availability Statement: No new data were created or analyzed in this study. Data sharing is not applicable to this article.

Acknowledgments: We thank our colleagues in the Eastern Bee Research Institute in Yunnan Agricultural University for their many helpful suggestions.

Conflicts of Interest: The authors declare no conflicts of interest.

References

1. Khalifa, S.A.M.; Elshafiey, E.H.; Shetaia, A.A.; El-Wahed, A.A.A.; Algethami, A.F.; Musharraf, S.G.; AlAjmi, M.F.; Zhao, C.; Masry, S.H.D.; Abdel-Daim, M.M.; et al. Overview of bee pollination and its economic value for crop production. *Insects* **2021**, *12*, 688. [CrossRef] [PubMed]
2. Hepburn, H.R.; Pirk, C.W.W.; Duangphakdee, O. *Honeybee Nests*; Springer: Berlin/Heidelberg, Germany, 2014.
3. Al-Fattah, M.A.A.-W.A.; Ibrahim, Y.Y.; Haggag, M.I. Some biological aspects of honey bee colonies in relation to the age of beeswax combs. *J. Apic. Res.* **2021**, *60*, 405–413.
4. Taha, E.-K.A.; Rakha, O.M.; Elnabawy, E.-S.M.; Hassan, M.M.; Shawer, D.M.B. Comb age significantly influences the productivity of the honeybee (*Apis mellifera*) colony. *J. King Saud Univ.-Sci.* **2021**, *33*, 101436.
5. Hepburn, H.R.; Kurstjens, S.P. The combs of honeybees as composite materials. *Apidologie* **1988**, *19*, 25–36.
6. Schmickl, T.; Crailsheim, K. Inner nest homeostasis in a changing environment with special emphasis on honey bee brood nursing and pollen supply. *Apidologie* **2004**, *35*, 249–263.
7. Brown, M.J.F.; Paxton, R.J. The conservation of bees: A global perspective. *Apidologie* **2009**, *40*, 410–416.
8. Feldhaar, H.; Otti, O. Pollutants and their interaction with diseases of social Hymenoptera. *Insects* **2020**, *11*, 153. [CrossRef]
9. Hassona, N.M.; El-Wahed, A.A.A. Heavy metal concentrations of beeswax (*Apis mellifera* L.) at different ages. *Bull. Environ. Contam. Toxicol.* **2023**, *111*, 26.
10. Ravoet, J.; Reybroeck, W.; de Graaf, D.C. Pesticides for apicultural and/or agricultural application found in Belgian honey bee wax combs. *Bull. Environ. Contam. Toxicol.* **2015**, *94*, 543–548.
11. Meng, Q.; Yang, S.; Huang, R.; Yang, L.; Xun, L.; Tian, Y.; Gong, X.; Wang, J.; Kuang, H.; Zhao, W.; et al. The significance of *Apis cerana cerana* (Hymenoptera: Apidae) gnawing off the old brood cells. *Apidologie* **2023**, *54*, 6.
12. Zhang, K.; Duan, H.; Karihaloo, B.L.; Wang, J. Hierarchical, multilayered cell walls reinforced by recycled silk cocoons enhance the structural integrity of honeybee combs. *Proc. Natl. Acad. Sci. USA* **2010**, *107*, 9502–9506. [CrossRef] [PubMed]
13. Dizaji, A.A.; Alishah, H.M.; Shaddel, A.-A.; Sis, N.M. Effects of comb wax age on the brood and honey product performance in honey bee. *Asian J. Anim. Vet. Adv.* **2008**, *3*, 51–53. [CrossRef]
14. Taha, E.-K.; Elsanat, S. Effect of combs age on honey production and its physical and chemical properties. In Proceedings of the 2nd International Conference of the Entomological Society of Egypt, Giza, Egypt, 8–11 December 2007; Springer: Berlin/Heidelberg, Germany, 2007; pp. 9–18.
15. Meng, Q.; Huang, R.; Li, H.; Gong, X.; Yue, D.; Jiang, W.; Tian, Y.; Dong, K. Analysis of comb-gnawing behavior in *Apis cerana cerana* (Hymenoptera: Apidae). *J. Insect Sci.* **2024**, *24*, 19. [CrossRef]
16. Hepburn, H.R.; Bernard, R.T.F.; Davidson, B.C.; Muller, W.J.; Lloyd, P.; Kurstjens, S.P.; Vincent, S.L. Synthesis and secretion of beeswax in honeybees. *Apidologie* **1991**, *22*, 21–36. [CrossRef]
17. Oeder, R.; Schwabe, D. The upward tilt of honeycomb cells increases the carrying capacity of the comb and is not to prevent the outflow of honey. *Apidologie* **2020**, *52*, 174–185. [CrossRef]
18. Ortis, G.; Frizzera, D.; Seffin, E.; Annoscia, D.; Nazzi, F. Honeybees use various criteria to select the site for performing the waggle dances on the comb. *Behav. Ecol. Sociobiol.* **2019**, *73*, 58. [CrossRef]
19. Hamdan, K. Natural supersedure of queens in honey bee colonies. *Bee World* **2010**, *87*, 52–54. [CrossRef]
20. Yang, S.; Deng, S.; Kuang, H.; Zhou, D.; Gong, X.; Dong, K. Evaluating and comparing the natural cell structure and dimensions of honey bee comb cells of Chinese bee, *Apis cerana cerana* (hymenoptera: Apidae) and Italian bee, *Apis mellifera ligustica* (hymenoptera: Apidae). *J. Insect Sci.* **2021**, *21*, 1–9. [CrossRef]
21. Winston, M.L. *The Biology of the Honey Bee*; Harvard University Press: London, UK, 1987.

22. Ma, D.; Huang, W. Apiculture in the new China. *Bee World* **1981**, *62*, 163–166.
23. Fratini, F.; Cilia, G.; Turchi, B.; Felicioli, A. Beeswax: A minireview of its antimicrobial activity and its application in medicine. *Asian Pac. J. Trop. Med.* **2016**, *9*, 839–843. [CrossRef]
24. Svečnjak, L.; Chesson, L.A.; Gallina, A.; Maia, M.; Martinello, M.; Mutinelli, F.; Muz, M.N.; Nunes, F.M.; Saucy, F.; Tipple, B.J.; et al. Standard methods for *Apis mellifera* beeswax research. *J. Apic. Res.* **2019**, *58*, 1–108. [CrossRef]
25. Hu, X.; Meng, Q.; Zhi, D.; Deng, S.; Zhao, W.; Tian, Y.; Gong, X.; Dong, K. Comb cell structure and morphological characteristics of the Chinese honey bee, *Apis cerana cerana* (Hymenoptera: Apidae), under successive generations. *J. Econ. Entomol.* **2023**, *116*, 352–358.
26. Bonvehi, J.S.; Orantes-Bermejo, F.J. Discoloration and adsorption of acaricides from beeswax. *J. Food Process Eng.* **2017**, *40*, e12344.
27. Maia, M.; Nunes, F.M. Authentication of beeswax (*Apis mellifera*) by high-temperature gas chromatography and chemometric analysis. *Food Chem.* **2013**, *136*, 961–968. [PubMed]
28. Bogdanov, S. Beeswax: Production, Properties, Composition, Control. In *Beeswax Book*; Bee Product Science Publishing: Muehlethurnen, Switzerland, 2016.
29. Jay, S.C. The development of honeybees in their cells. *J. Apic. Res.* **1963**, *2*, 117–134.
30. Jay, S.C. The cocoon of the honey bee, *Apis mellifera* L. *Can. Entomol.* **1964**, *96*, 784–792. [CrossRef]
31. Klein, B.A.; Busby, M.K. Slumber in a cell: Honeycomb used by honey bees for food, brood, heating... and sleeping. *PeerJ* **2020**, *5*, e9583.
32. Hepburn, H.R.; Muerrle, T.; Radloff, S.E. The cell bases of honeybee combs. *Apidologie* **2007**, *38*, 268–271.
33. Graham, J.M. *The Hive and the Honey Bee*; Dadant & Sons: Hamilton, IL, USA, 1992.
34. Elnabawy, E.M.; Mousa, K.M.; Ueno, T.; Shawer, M.B. Impact of different comb age on morphological and biological characteristics of honeybee workers (*Apis Mellifera* L.). *J.-Fac. Agric. Kyushu Univ.* **2020**, *65*, 277–282.
35. Ruttner, F. *Biogeography and Taxonomy of Honeybees*; Springer: Berlin/Heidelberg, Germany, 1988.
36. Smith, F.G. The races of honeybees in Africa. *Bee World* **1961**, *42*, 255–260.
37. Dutton, R.W.; Ruttner, F.; Berkeley, A.; Manley, M.J.D. Observations on the morphology, relationships and ecology of *Apis Mellifera* of Oman. *J. Apic. Res.* **1981**, *20*, 201–214.
38. Khan, M.S.; Kaushik, H.D.; Rohilla, H.R. Nesting behaviour of the rock bee, *Apis dorsata* at Hisar, India: II-height and directional preferences for comb building. *Indian Bee J.* **2007**, *69*, 8–12.
39. Starr, C.K. *Encyclopedia of Social Insects*; Springer Nature Switzerland AG: Cham, Switzerland, 2021.
40. Free, J.B. Biology and behaviour of the honeybee *Apis Florea*, and possibilities for beekeeping. *Bee World* **1981**, *62*, 46–59.
41. Zhang, L.; Shao, L.; Raza, M.F.; Han, R.; Li, W. The effect of comb cell size on the development of *Apis mellifera* drones. *Life* **2024**, *14*, 222. [CrossRef] [PubMed]
42. Li, H.; Deng, S.; Meng, Q.; Yang, S.; Jiang, W.; Zhao, W.; Tian, Y.; Dong, K.; Gong, X. Impact of *Apis cerana cerana* (Hymenoptera: Apidae) body size on newly built cell size. *J. Asia-Pac. Entomol.* **2024**, *27*, 102277.
43. Volosevich, A.P.; Kulzhinskaya, K.P. Improved rearing of bees by changing the feeding conditions. *Bee World* **1957**, *38*, 101–112.
44. Abdellatif, M. Comb cell size and its effect on the body weight of the worker bee, *Apis mellifera* L. *Am. Bee J.* **1965**, *105*, 86–87.
45. Berry, J.A.; Delaplane, K.S. Effects of comb age on honey bee colony growth and brood survivorship. *J. Apic. Res.* **2001**, *40*, 3–8.
46. Sipos, T.; DonkÓ, T.; CsÓka, Á.; Kiss, T.; Keszthelyi, S. Comparative micro-computed tomographic analysis of the structure of brood cells and its effect on the development of the pupae of honey bee (*Apis mellifera*). *Eur. J. Entomol.* **2023**, *120*, 9–14.
47. AL-Kahtani, S.N. Morphometric characteristics of carniolan honeybee workers in relation to age of comb. *Sci. J. King Faisal Univ. (Basic Appl. Sci.)* **2018**, *19*, 47–54.
48. Nazzi, F. The hexagonal shape of the honeycomb cells depends on the construction behavior of bees. *Sci. Rep.* **2016**, *6*, 28341.
49. AlKahtani, S.N.; Taha, E.A. Morphometric study of Yemeni (*Apis mellifera jemenitica*) and Carniolan (*A. m. carnica*) honeybee workers in Saudi Arabia. *PLoS ONE* **2021**, *16*, e0247262.
50. Dizaji, A.A.; Sis, N.M.; Tili, A.S.; Golshani, A.A. Effect of comb wax age on brood performance or population of honey bee, *Apis melifera*. *Agroecol. J.* **2009**, *5*, 89–94.
51. Taha, E.; Al-Kahtani, S.N. The relationship between comb age and performance of honey bee (*Apis mellifera*) colonies. *Saudi J. Biol. Sci.* **2019**, *27*, 30–34. [CrossRef] [PubMed]
52. Shapla, U.M.; Solayman, M.; Alam, N.; Khalil, M.I.; Gan, S.H. 5-Hydroxymethylfurfural (HMF) levels in honey and other food products: Effects on bees and human health. *Chem. Cent. J.* **2018**, *12*, 35. [CrossRef]
53. Miquelena, E.; Piccirillo, G.; Rodriguez, B. Influence of comb age on some physical-chemical properties of honeys produced by African bees. *Rev. De La Fac. De Agron.* **2010**, *27*, 626–638.
54. Taha, E.-K.A.; Manosur, H.M.; Shawer, M.B. The relationship between comb age and the amounts of mineral elements in honey and wax. *J. Apic. Res.* **2010**, *49*, 202–207.
55. Fröhlich, B.; Tautz, J.; Riederer, M. Chemometric classification of comb and cuticular waxes of the honeybee *Apis mellifera carnica*. *J. Chem. Ecol.* **2000**, *26*, 123–137. [CrossRef]

56. Zhang, S.; Ma, J.; Zhang, L.; Sun, Z.; Zhao, Z.; Khan, N. Does adoption of honeybee pollination promote the economic value of kiwifruit farmers? Evidence from China. *Int. J. Environ. Res. Public Health* **2022**, *19*, 8305. [CrossRef]
57. Eeraerts, M.; DeVetter, L.W.; Batáry, P.; Ternest, J.J.; Mallinger, R.; Arrington, M.; Benjamin, F.E.; Blaauw, B.R.; Campbell, J.W.; Cavigliasso, P.; et al. Synthesis of highbush blueberry pollination research reveals region-specific differences in the contributions of honeybees and wild bees. *J. Appl. Ecol.* **2023**, *60*, 2528–2539. [CrossRef]
58. Nowak, A.; Nowak, I. Review of harmful chemical pollutants of environmental origin in honey and bee products. *Crit. Rev. Food Sci. Nutr.* **2023**, *63*, 5094–5116. [CrossRef]
59. Martinello, M.; Manzinello, C.; Dainese, N.; Giuliato, I.; Gallina, A.; Mutinelli, F. The honey bee: An active biosampler of environmental pollution and a possible warning biomarker for human health. *Appl. Sci.* **2021**, *11*, 6481. [CrossRef]
60. Meng, Q.; Huang, R.; Yang, S.; Li, H.; Yue, D.; Gong, X.; Zhao, W.; Tian, Y.; Dong, K. Impact of brood cell cocoons on metal accumulation and CYP450 detoxification gene expression in *Apis cerana cerana*. *Toxics* **2024**, *12*, 131. [CrossRef]
61. Herrero-Latorre, C.; Barciela-García, J.; García-Martín, S.; Peña-Crecente, R.M. The use of honeybees and honey as environmental bioindicators for metals and radionuclides: A review. *Environ. Rev.* **2017**, *25*, 463–480.
62. Bargańska, Ż.; Ślebioda, M.; Namieśnik, J. Honey bees and their products: Bioindicators of environmental contamination. *Crit. Rev. Environ. Sci. Technol.* **2016**, *46*, 235–248.
63. Gajger, I.T.; Kosanović, M.; Oreščanin, V.; Kos, S.; Bilandžić, N. Mineral content in honeybee wax combs as a measurement of the impact of environmental factors. *Bull. Environ. Contam. Toxicol.* **2019**, *103*, 697–703. [PubMed]
64. Scivicco, M.; Nolasco, A.; Esposito, L.; Ariano, A.; Squillante, J.; Esposito, F.; Cirillo, T.; Severino, L. Effects of COVID-19 pandemic lockdown and environmental pollution assessment in Campania region (Italy) through the analysis of heavy metals in honeybees. *Environ. Pollut.* **2022**, *307*, 119504. [CrossRef]
65. Gizaw, G.; Kim, Y.; Moon, K.; Choi, J.B.; Kim, Y.H.; Park, J.K. Effect of environmental heavy metals on the expression of detoxification-related genes in honey bee *Apis mellifera*. *Apidologie* **2020**, *51*, 664–674. [CrossRef]
66. Yang, Y.; Wu, Y.; Long, H.; Ma, X.; Shariati, K.; Webb, J.; Guo, L.; Pan, Y.; Ma, M.; Deng, C.; et al. Global honeybee health decline factors and potential conservation techniques. *Food Secur.* **2023**, *15*, 855–875. [CrossRef]
67. Alkassab, A.T.; Bischoff, G.; Thorbahn, D.; Frommberger, M.; Pistorius, J. Transfer of xenobiotics from contaminated beeswax into different bee matrices under field conditions and the related exposure probability. *Chemosphere* **2022**, *307*, 135615.
68. Wallner, K. Varroacides and their residues in bee products. *Apidologie* **1999**, *30*, 235–248. [CrossRef]
69. Medici, S.K.; Castro, A.; Sarlo, E.G.; Marioli, J.M.; Eguaras, M.J. The concentration effect of selected acaricides present in beeswax foundation on the survival of *Apis mellifera* colonies. *J. Apic. Res.* **2012**, *51*, 164–168. [CrossRef]
70. Таранов, Г.Ф. Биология Пчелиной Семьи; 1961, Сельхозиздат: Госсельхозиздат. 1-331. Available online: https://tochok.info/files/file/121-%D1%82%D0%B0%D1%80%D0%B0%D0%BD%D0%BE%D0%B2-%D0%B3-%D1%84-%D0%B1%D0%B8%D0%BE%D0%BB%D0%BE%D0%B3%D0%B8%D1%8F-%D0%BF%D1%87%D0%B5%D0%BB%D0%B8%D0%BD%D0%BE%D0%B9-%D1%81%D0%B5%D0%BC%D1%8C%D0%B8-1961/ (accessed on 25 March 2025).
71. Gong, Y.; Diao, Q. Current knowledge of detoxification mechanisms of xenobiotic in honey bees. *Ecotoxicology* **2017**, *26*, 1–12.
72. Zhu, M.; Zhang, W.; Liu, F.; Chen, X.; Li, H.; Xu, B. Characterization of an *Apis cerana cerana* cytochrome P450 gene (*AccCYP336A1*) and its roles in oxidative stresses responses. *Gene* **2016**, *584*, 120–128.
73. Rand, E.E.d.; Smit, S.; Beukes, M.; Apostolides, Z.; Pirk, C.W.W.; Nicolson, S.W. Detoxification mechanisms of honey bees (*Apis mellifera*) resulting in tolerance of dietary nicotine. *Sci. Rep.* **2015**, *5*, 11779.
74. Amezian, D.; Nauen, R.; Le Goff, G. Transcriptional regulation of xenobiotic detoxification genes in insects—An overview. *Pestic. Biochem. Physiol.* **2021**, *174*, 104822.
75. Ellis, J.D.; Graham, J.R.; Mortensen, A. Standard methods for wax moth research. *J. Apic. Res.* **2013**, *52*, 1–17.
76. Kwadha, C.A.; Ong'amo, G.O.; Ndegwa, P.N.; Raina, S.K.; Fombong, A.T. The biology and control of the greater wax moth, *Galleria mellonella*. *Insects* **2017**, *8*, 61. [CrossRef]
77. Navarro-Hortal, M.D.; Orantes-Bermejo, F.J.; Sánchez-González, C.; Varela-López, A.; Giampieri, F.; Torres Fernández-Piñar, C.; Serra-Bonvehí, J.; Forbes-Hernández, T.Y.; Reboredo-Rodríguez, P.; Llopis, J.; et al. Industrial-scale decontamination procedure effects on the content of acaricides, heavy metals and antioxidant capacity of beeswax. *Molecules* **2019**, *24*, 1518. [CrossRef]
78. Luna, A.; Alonso, R.; Cutillas, V.M.; Ferrer, C.M.; Gómez-Ramos, M.J.; Hernando, D.; Valverde, A.; Flores, J.M.; Fernández-Alba, A.R.; Fernández-Alba, A.R. Removal of pesticide residues from beeswax using a methanol extraction-based procedure: A pilot-scale study. *Environ. Technol. Innov.* **2021**, *23*, 101606.

insects

MDPI

Review

An Overview of the Nutritional Requirements of Honey Bees (*Apis mellifera* Linnaeus, 1758)

Leticia S. Ansaloni *, Janja Kristl, Caio E. C. Domingues and Aleš Gregorc

Faculty of Agriculture and Life Sciences, University of Maribor, Pivola 10, 2311 Hoče, Slovenia; janja.kristl@um.si (J.K.); cecdomingues@gmail.com (C.E.C.D.); ales.gregorc@um.si (A.G.)
* Correspondence: leticia.salvioni1@um.si

Simple Summary: It is well documented in the literature that bees have been facing numerous stressors in recent decades. Although honey bees are well managed worldwide, they have not been unaffected by these challenges. Honey bees are important for pollination in various crops, for the beekeeping sector and for the natural habitats as well. However, to perform this role effectively, they require proper nutrition to support their healthy growth and that of the colony. In the current review, we broadly discuss the basic nutritional needs of honey bees, such as the consumption of proteins, carbohydrates, lipids, amino acids, vitamins, minerals, water, and other essential components. Additionally, we also discuss the use of nutritional supplements aimed at supporting the health and strengthening of the colony, and how, despite these efforts, inadequate nutrition can still be a problem for honey bee colonies.

Abstract: Honey bees are known for their wide global distribution, their ease of handling, and their economic and ecological value. However, they are often exposed to a wide variety of stress factors. Therefore, it is essential for beekeepers to maintain healthy bee colonies. In this context, a balanced diet is recommended to support the growth of strong and healthy honey bee colonies. The purpose of this review is therefore to provide an overview of the nutritional requirements of *Apis mellifera* and their importance for the maintenance of healthy bee colonies. An adequate diet includes the consumption of sufficient amounts of proteins, carbohydrates, lipids, amino acids, vitamins, minerals, water, and essential sterols, and a diet based on multi-floral pollen is desirable. However, when honey bee colonies are located near agroecosystems with lower resource diversity, both brood rearing and colony longevity may decrease, making them more susceptible to parasites and diseases. On the other hand, efforts have been made to improve the health of honey bee colonies with the help of nutritional supplements consisting of a variety of components. Nevertheless, studies have shown that even with these supplements, a lack of nutrients can still be an issue for honey bee colonies. Furthermore, future research should focus on identifying nutritional supplements that can better replicate natural diet diversity and assessing long-term effects on honey bee colony resilience, especially in low-flowering areas. This review discusses the interaction between nutrient requirements and the effects of supplements on colony health.

Keywords: development; health; nectar; nutrition; pollen

1. Introduction

Pollination stands out as one of the most essential ecosystem services. According to Ollerton et al. [1], it is estimated that approximately 87.5% of all flowering plant species depend on animal pollination. Bees provide valuable pollination services to a wide range of crops and native flora worldwide [2–4]. Among the diversity of bee species, with over 20,000 already described worldwide [5], the honey bee (*Apis mellifera* Linnaeus, 1758) distinguishes itself globally for its substantial economic and ecological value, derived from the essential role it plays in pollination [6]. This species holds significant economic importance by not only increasing production and quality in agricultural fields but also through the commercialization of bee products like apitoxin (bee venom), honey, propolis, royal jelly, and wax [3,7].

According to Ilyasov et al. [8], although there are differences in taxonomic revisions, thirty-three distinct honey bee subspecies are currently found in Europe, Africa, and Western Asia and the Middle East. Moreover, due to the Africanization process resulting from the intercross of European and African subspecies, the hybrid Africanized *A. mellifera* has also spread throughout the Americas [9,10]. Depending on the subspecies, honey bees may exhibit less defensive behavior and an enhanced ability to adapt to colder climates, as observed in *A. m. carnica* [11], or more defensive behavior, as found in *A. m. scutellata* [12], as well as better adaptability to warm environments, as described for *A. m. ligustica* [13]. Nevertheless, the honey bees have a strong foraging orientation and exhibit high performance in honey production [14,15].

Many studies have been conducted within breeding programs for *A. mellifera* to improve desirable honey bee traits, aiming to obtain healthier and more productive colonies [11,16–20]. These efforts also include enhancing resistance to Varroa [21–23], and screening for the presence of viruses during queen rearing [24–26].

However, honey bee colony mortality has impacted beekeepers worldwide over the past decades [27], calling for effective mitigation strategies. Goulson et al. [28] highlighted multiple factors associated with the decline in bees, such as habitat loss, parasites and disease, pesticides, monotonous diets, shipping fever, competition, and climate change. It is currently known that the health risks for honey bees increase when two or more stressors are combined [29]. Nowadays, it is crucial for farmers and beekeepers worldwide to maintain healthy honey bee colonies, especially since these colonies are often exposed to stressors and are frequently located near agroecosystems and natural forest areas [30].

Winkler et al. [31] reported that global land use has increased more than previously estimated in recent decades, which may be linked to global trade in agricultural production. Similarly, large-scale monoculture has negatively impacted apiculture by reducing honey yield, as described by de Groot et al. [32]. As mentioned before, monotonous diets are one of the stress factors for bees, and honey bees are highly dependent on the availability and diversity of floral resources, which are crucial for the development and survival of their colonies [33]. This combination of diverse floral sources is important because, depending on the species, these sources can vary in protein, amino acid, lipid, carbohydrates, vitamins, and other nutritional contents [34]. Concerns about honey bees receiving an inadequate diet have driven the development of protein supplements, and efforts have been made to study the effectiveness of these supplements on colony health [35–37]. Brodschneider and Crailsheim [34] suggested that a balanced diet is necessary to maintain well-fed and healthy colonies. Similarly, queens, workers, and drones have different physiology and nutritional demands during their development and throughout their lives [38], which further highlights the importance of nutritional diversity.

Based on the aforementioned considerations and the importance of nutrition for healthy colonies, this study aims to provide an overview of the diet and nutritional requirements of *A. mellifera*, as well as how these factors influence the development of larvae and adult, including the use of nutritional supplements.

2. Nutritional Requirements for Honey Bees

A. mellifera is the most frequently managed bee species globally [39] and is found on all continents except Antarctica. The honey bee has been used as a model organism for the study of various topics due to the extensive knowledge of its biology [40]. In this regard, research in bee nutrition has made significant progress over time, allowing a better understanding of the specific nutritional requirements of *A. mellifera* [34,41]. The essential nutrients required for the growth, development, and survival of honey bee colonies are highlighted below.

2.1. Proteins

Pollen is the main source of protein for honey bees, which they collect from a wide variety of plants [3]. Studies have shown that the protein content of pollen varies depending on the plant species [42]. Radev [43] determined the crude protein content in mixed pollen species collected during different harvesting periods (April to September) in Bulgaria and found a range of 11.5% to 27.4%, while a literature review by Roulston et al. [44] found a wider range of protein content from 2.5% to 61% in the pollen of 377 plant species from 93 plant families. In this regard, the quality and diversity of pollen are important for honey bee physiology and health [33], as the protein content and amino acid profile can vary, and a lack of pollen diversity can impair development and leave the colony more susceptible to pathogens and diseases [45]. A diet consisting of multi-flower pollen is preferable [46], as balanced nutrition more effectively ensures the proper growth, development, and overall health of honey bee colonies.

The amount of pollen collected by a honey bee colony can be influenced by different factors, including climatic conditions, season, colony size, health status, and the availability and diversity of floral sources [47,48], which also affect protein levels in the colony. Wille et al. [49] described a range of 10–26 kg of pollen collected per year, whereas Avni et al. [50] reported 16.8 kg of pollen per colony as an overall annual mean, with the variations depending on sites and seasons. However, higher values, with pollen collection reaching up to 55 kg per year, have already been reported in the literature [51]. The nutritional value of pollen is a valuable source of amino acids, lipids, carbohydrates, minerals, vitamins, and essential sterols [52]. These components are closely linked to the growth, reproduction, and productivity of *A. mellifera* colonies [53]. According to Radev [54], honey bee colony reproduction increases in the presence of pollen with high protein content, specifically over 21 and 27%. In contrast, if pollen access is reduced, brood rearing and colony lifespan may decrease [55].

The protein requirements of adult bees and honey bees' larvae are different. A study by Crailsheim et al. [56] showed that the daily pollen consumption of an adult honey bee was 3.4–5.4 mg, depending on age, while the entire colony consumed between 13.4 and 17.8 kg of pollen annually. In contrast, Hrassnigg and Crailsheim [38] emphasized that larvae require a higher protein intake and need to consume 25–37.5 mg per reared larva (125–187.5 mg of pollen with a protein content of about 20%). According to Li et al. [57], a higher protein content in the larval diet can increase worker longevity, possibly due to the upregulation of genes related to antioxidant activity. A pollen-based diet is also associated with lower sensitivity to pesticides [58]. In addition, honey bees use pollen to make bee bread, a food consisting of a mixture of pollen, regurgitated nectar, and enzymes secreted by their glands [59]. Bee bread provides energy, has a higher nutritional value than pollen, and contains bioactive compounds such as phenolics and flavonoids, which exhibit antioxidant and antimicrobial activity [59,60]. According to Kieliszek et al. [59], it is more digestible for bees due to the presence of amino acids and serves as the main food source for larvae and young bees. However, the potential of bee bread has not yet been explored, and there is little relevant literature in comparison to protein feeds.

The use of protein-supplemented diets in honey bee colonies has increased among beekeepers worldwide in recent years [37,61]. DeGrandi-Hoffman et al. [62] observed that a commercial diet with 16.5% protein and 66% carbohydrate can stimulate the production of more brood. According to Zheng et al. [63], the reproduction rate of honey bees was maximized when they were fed with a crude protein content of 29.5–34.0%. Liquid protein supplementation (Hydroamin 30®—Bioal diluted at 4% in one liter of sucrose–water syrup 2:1 m/v) improved bee health and decreased post-winter mortality, as highlighted by García-Vicente et al. [64]. Given these findings, beekeepers should consider adding protein supplements to the diet during pollen shortages. On the other hand, DeGrandi-Hoffman et al. [65] demonstrated that natural forage was more effective than protein supplements in reducing pathogen load and resulted in higher over-winter survival, which was related to the greater digestion of pollen proteins. The authors suggest that the protein supplements were less digestible than pollen because the protein sources (soy, barley flour, and eggs) are not part of the bees' diet. However, additional research is required to evaluate the effects of supplemented protein diets. Furthermore, bees are no longer finding the same resources that were available to them in previous years, which is why fructose and sucrose syrups are being more widely used. Studies have reported the impacts of these changes on bee nutrition and health [66,67].

2.2. Carbohydrates

Nectar is the most important source of carbohydrates (sucrose, glucose, and fructose) for honey bees, and flowers are their primary natural source [34,68]. Similar to the protein content in pollen, the carbohydrate ratio in nectar also varies depending on the plant species and its structural characteristics (e.g., flower shape and nectary structure) [69,70]. Additionally, honeydew, a sugary excretion produced by many different sap-feeding insects, serves as another important source of carbohydrates for honey bees [71].

Carbohydrates are an essential source of energy, providing the necessary fuel for foraging and internal colony activities [72]. According to a systematic review performed by Pamminger et al. [73], the average sugar concentration in nectar from crops, weeds, and wild plants worldwide is 40%, with the optimal sugar concentration suggested to range from 35% to 65%, based on uptake measurements and theoretical considerations by Kim et al. [74]. The nutritional value of nectar is based on three primary sugars: sucrose, glucose, and fructose [69]. Other sugars occur in lower concentrations or are difficult to break down [75]. Additionally, some nectars can be toxic to honey bee if ingested [76], such as from *Sophora microphylla*, which contains alkaloids (e.g., cytisine) [77], and *Ochroma lagopus*, known for its alkaloid content (e.g., atropine and scopolamine) [78]. These alkaloids are known for their toxic effects on the nervous system, physiology, and behavior of honey bees. In addition, among toxic nectars, it is important to highlight those from plants that contain pyrrolizidine alkaloids (PAs), the toxicity of which to adult honey bees and brood has already been reported [79–82]. This has raised concerns due to its potential risks to human health, as PAs are currently under examination by the European Food Safety Authority (EFSA) because of their toxicity [83–85].

Paoli et al. [86] demonstrated that the nutritional demand for carbohydrates increases depending on the honey bees' age, with foragers (older bees) consuming 60% more than younger bees aged 0–7 and 8–14 days. The authors estimated that the carbohydrate demand increases from two to five times for foragers. According to Couvillon et al. [87], the foraging distances of honey bees vary depending on the month and the type of forage available. Based on the analysis of 4562 waggle dances decoded from nectar foragers, the average foraging distance was 1408 m, with the maximum distance exceeding 5000 m. These long distances demand a high energy expenditure. Additionally, larvae also need carbohydrates for the development, with the estimated sugar requirement being 59.4 mg for workers (5 days) and 98.2 mg for drones (6.5 days) [51].

Nectar is important as the main source for making honey [34]. Honey bees process the nectar by adding enzymes from their salivary glands to break down complex sugars and reducing large amounts of water to prevent fermentation [75]. The processed nectar is stored in combs and serves as an essential source of food during floral resource scarcity [52]. According to Silva et al. [88], honey is primarily a blend of sugars, water, proteins, vitamins, organic acids, minerals, pigments, phenolic compounds, and volatile compounds. The main carbohydrates present in honey are fructose (38.2%) and glucose (31.3%) on average [89], although sucrose, maltose, and other sugars can also be found [88].

The use of carbohydrate supplementation is also a common practice in beekeeping [61,75,90–92]. Sammataro and Weiss [75] highlighted that colonies fed with sucrose syrup built more honeycomb, had a higher average mass of bees, and showed increased spring brood production compared to colonies fed with high-fructose corn syrup, which is commonly used as a source of bee feed. On the other hand, honey bees exhibited sublethal effects on stress proteins, such as superoxide dismutase, suggesting an impaired response to oxidative stress when fed with sucrose syrup and high-fructose corn syrup [91]. Wheeler et al. [90] suggested that a diet of 50% (w/v) honey provides more nutritional components compared to 50% high-fructose corn syrup or 50% sucrose, which are widely used food sources in apiculture. According to Frizzera et al. [93], since sugar supplementation has become an increasingly common practice among beekeepers, it is necessary to be aware about the potential side effects on honey bee health, including a decrease in bee survival.

2.3. Lipids

Honey bees obtain lipids from the pollen they collect [52]. According to the findings of Dobson [94], nonpolar lipids may be present in the pollenkitt, while polar lipids are found in the inner pollen fraction. Both types of lipids are important for honey bees. The lipid content and composition in pollen varies depending on the plant species [44,95]. Roulston and Cane [44] compiled published data for 62 plant species, with the lipid content ranging from 0.8% in *Eucalyptus marginata* (Myrtaceae) to 18.9% in *Taraxacum officinale* (Asteraceae).

Lipids are essential for physiological processes related to energy, metabolism, and regulation [96]. They serve as a significant energy source for honey bees, store energy reserves, and are key components of cell membrane structures [96,97]. Lipids are primarily stored in the fat body, with larvae having a greater amount of lipids compared to adult honey bees [98]. However, studies have also shown that young nurse bees store a large amount of lipids in their abdomens [99,100]. Conversely, the amount of abdominal lipids decreases with age [101], which could be related to foraging activity.

In addition, fatty acids are also important for honey bees and contribute to antimicrobial properties, cognitive functions, and reproduction [52,100,102]. Palmitic acid, linoleic acid (omega-6), and alpha-linolenic acid (omega-3) are the most commonly found fatty acids in pollen [102], and together with oleic acid, they are also present in the body lipids of drones, workers, and queens [103].

Compared to other types of dietary supplementations, there are few studies on the use of lipid diets. Nonetheless, research has demonstrated that brood production in honey bees increases when lipid extracts (e.g., pollen lipid fraction) are included in their diet [104,105]. According to Arien et al. [105], brood rearing was best with an 8% lipid diet, while survival was highest with a 4% lipid diet and an omega-6:3 ratio of 1. However, further studies are needed to understand the long-term effects of feeding lipid diets.

2.4. Amino Acids and Vitamins

Amino acids also contribute significantly to the nutritional needs of honey bees, with key functions in their metabolism, including immune response [106], phagostimulatory effects [107], and learning and memory [108], as well as serving as neurotransmitters and contributing to other essential processes [109]. Pollen is a valuable source of amino acids, especially exogenous ones [110]. According to De Groot [111], the amino acids arginine, histidine, lysine, phenylalanine, tryptophan, leucine, isoleucine, threonine, and valine are essential for the development of adult honey bee workers.

In addition to amino acids, vitamins are also essential nutrients for honey bees and can be found in pollen, nectar, honey, royal jelly, and other food sources [52]. The concentrations of various vitamins, including thiamine, riboflavin, niacin, pantothenic acid, pyridoxine, folic acid, cobalamin, biotin, vitamin E, and ascorbic acid, vary in pollen, royal jelly, and honey [112]. A study conducted by Elsayeh et al. [112] demonstrated that B-vitamins can play an important role in regulating the uptake of macronutrients, such as carbohydrates and essential amino acids in honey bees.

Royal jelly (RJ) is a highly nutritious secretion produced by young nurse worker bees through their hypopharyngeal and mandibular glands [113]. It serves as an essential food source for the development of the larvae, especially the queens [68]. The composition of RJ includes water, proteins, free amino acids, carbohydrates, lipids, and minerals [114]. According to Howe et al. [115], RJ contains 17 amino acids and five unidentified ninhydrin-positive compounds. On the other hand, Liming et al. [116] determined 26 amino acids in RJ. This intricate balance of vitamins and amino acids in RJ, along with other food sources, contributes to the overall health and development of honey bees.

Both amino acids and vitamins have been used as food supplements for honey bees [117–119]. Glavinic et al. [118] demonstrated that newly emerged worker bees fed a diet supplemented with a combination of vitamins, minerals, and amino acids had a significantly reduced number of Nosema spores compared to the control, particularly on day 12 post-infection. This effect was observed in the context of immunosuppression induced by the microsporidium *Nosema ceranae*. Similarly, Stanimirović et al. [119] demonstrated a significant and consistent increase in the hygienic behavior of honey bees when they were fed with the same diet mentioned above, which can help the colony resist microsporidial and viral infections.

2.5. Water

In addition to the nutrient requirements described above, water is also essential for the survival of honey bees. According to Nicolson [120], bees use water not only to maintain osmotic homeostasis but also for several other tasks within the hive, such as cooling the brood area, regulating the hive temperature, and feeding the larval brood, due to the high water content in RJ. Moreover, water can also provide minerals to bees [121]. Mineral nutrients such as sodium (Na), magnesium (Mg), and potassium (K) are essential to meet the nutritional needs of bees [122,123]. According to Khan et al. [124], honey bee colonies showed a preference for sodium chloride (NaCl), potassium chloride (KCl), and magnesium chloride ($MgCl_2$) over deionized water in both the summer and winter seasons.

A study conducted by Filipiak et al. [121] suggested that Na deficiencies in honey bee diets must be supplemented using "dirty water", which is rich in salts [123]. Similarly, Cairns et al. [125] highlighted the importance of stoichiometrically balanced diets for honey bees, suggesting that "dirty water" may be a good source of mixed minerals. However, it is important to ensure that any supplementary sources do not harm the bees. Further studies are needed to better understand honey bee water foraging preferences.

3. Conclusions

Considering all the nutritional requirements of *A. mellifera*, including the need for a well-balanced diet, along with the challenges posed by proximity to agricultural crops and exposure to various stressors, maintaining healthy honey bee colonies has become more challenging for beekeepers. As a result, in response to the demand for stronger and more productive colonies, the use of dietary supplementation in beekeeping practices is becoming more common worldwide, especially because the availability of resources is not always possible. Additionally, decisions regarding food supplementation factors, such as landscape use, climatic conditions, seasonal variations, and the presence of environmental stressors, should be considered, to optimize colony health and productivity. Further research needs to be conducted in order to evaluate the long-term effectiveness of food supplementation on honey bee colonies.

Author Contributions: Conceptualization, L.S.A., J.K., C.E.C.D., and A.G.; investigation, L.S.A., J.K., C.E.C.D., and A.G.; writing—original draft preparation, L.S.A., J.K., C.E.C.D., and A.G.; writing—review and editing, L.S.A., J.K., C.E.C.D., and A.G.; supervision, J.K., and A.G.; funding acquisition, L.S.A., and A.G. All authors have read and agreed to the published version of the manuscript.

Funding: This research was supported by the Public Agency for Research Activity of the Republic of Slovenia under the Contract on Stable Financing of Scientific Research Activity for the Contract Period 2022-2027, no. S-ZRD/22-27/552, and the Research core funding No. P4-164; Research for improvement of safe food and health, by the project CRISPR-B number N4-0192.

Institutional Review Board Statement: Not applicable.

Data Availability Statement: Not applicable.

Acknowledgments: The authors thank the anonymous reviewers for their contributions to this version of the review article.

Conflicts of Interest: The authors declare no conflicts of interest.

References

1. Ollerton, J.; Winfree, R.; Tarrant, S. How many flowering plants are pollinates by animals? *Oikos* **2011**, *120*, 321–326. [CrossRef]
2. Ollerton, J. Pollinator diversity: Distribution, ecological function, and conservation. *Annu. Rev. Ecol. Evol. Syst.* **2017**, *48*, 353–376. [CrossRef]
3. Hung, K.L.J.; Kingston, J.M.; Albrecht, M.; Holway, D.A.; Kohn, J.R. The worldwide importance of honey bees as pollinators in natural habitats. *Proc. R Soc. Lond. B Biol. Sci.* **2018**, *285*, 20172140. [CrossRef] [PubMed]
4. Basualdo, M.; Cavigliasso, P.; de Avila, R.S., Jr.; Aldea-Sánchez, P.; Correa-Benítez, A.; Harms, J.M.; Ramos, A.K.; Rojas-Bravo, V.; Salvarrey, S. Current status and economic value of insect-pollinated dependent crops in Latin America. *Ecol. Econ.* **2022**, *196*, 107395. [CrossRef]
5. Orr, M.C.; Hughes, A.C.; Chesters, D.; Pickering, J.; Zhu, C.D.; Ascher, J.S. Global patterns and drivers of bee distribution. *Curr. Biol.* **2021**, *31*, 451–458. [CrossRef]
6. Paudel, Y.P.; Mackereth, R.; Hanley, R.; Qin, W. Honey bees (*Apis mellifera* L.) and pollination issues: Current status, impacts, and potential drivers of decline. *J. Agric. Sci.* **2015**, *7*, 93–109. [CrossRef]
7. Morse, R.A.; Calderone, N.W. The value of honey bees as pollinators of U.S. Crops in 2000. *Bee Cult.* **2000**, *128*, 1–15.
8. Ilyasov, R.A.; Lee, M.L.; Takahashi, J.I.; Kwon, H.W.; Nikolenko, A.G. A revision of subspecies structure of western honey bee *Apis mellifera*. *Saudi J. Biol. Sci.* **2020**, *27*, 3615–3621. [CrossRef]
9. Branchiccela, B.; Aguirre, C.; Parra, G.; Estay, P.; Zunino, P.; Antúnez, K. Genetic changes in *Apis mellifera* after 40 years of Africanization. *Apidologie* **2014**, *45*, 752–756. [CrossRef]
10. Porrini, L.P.; Quintana, S.; Brasesco, C.; Porrini, M.P.; Garrido, P.M.; Eguaras, M.J.; Müller, F.; Fernandez Iriarte, P. Southern limit of Africanized honey bees in Argentina inferred by mtDNA and wing geometric morphometric analysis. *J. Apic. Res.* **2020**, *59*, 648–657. [CrossRef]
11. Hoppe, A.; Du, M.; Bernstein, R.; Tiesler, F.-K.; Kärcher, M.; Bienefeld, K. Substantial genetic progress in the international *Apis mellifera carnica* population since the implementation of genetic evaluation. *Insects* **2020**, *11*, 768. [CrossRef] [PubMed]

12. Tarekegn, A.; Faji, M.; Abebe, A. Production performance and various important behaviors performed by the *Apis mellifera scutellata* bee race. *Uludağ Arıcılık Derg.* **2022**, *22*, 211–226. [CrossRef]
13. Kovac, H.; Käfer, H.; Stabentheiner, A.; Costa, C. Metabolism and upper thermal limits of *Apis mellifera carnica* and *A. m. ligustica*. *Apidologie* **2014**, *45*, 664–677. [CrossRef] [PubMed]
14. Beekman, M.; Ratnieks, F.L.W. Long-range foraging by the honey-bee, *Apis mellifera* L. *Funct. Ecol.* **2000**, *14*, 490–496. [CrossRef]
15. Olate-Olave, V.R.; Verde, M.; Vallejos, L.; Perez Raymonda, L.; Cortese, M.C.; Doorn, M. Bee health and productivity in *Apis mellifera*, a consequence of multiple factors. *Vet. Sci.* **2021**, *8*, 76. [CrossRef]
16. Gregorc, A.; Lokar, V. Selection criteria in an apiary of Carniolan honey bee (*Apis mellifera carnica*) colonies for queen rearing. *J. Cent. Eur. Agric.* **2010**, *11*, 401–408. [CrossRef]
17. Hatjina, F.; Bieńkowska, M.; Charistos, L.; Chlebo, R.; Costa, C.; Dražić, M.M.; Filipi, J.; Gregorc, A.; Ivanova, E.N.; Kezić, N.; et al. A review of methods used in some European countries for assessing the quality of honey bee queens through their physical characters and the performance of their colonies. *J. Apic. Res.* **2014**, *53*, 337–363. [CrossRef]
18. Gregorc, A.; Škerl, M.I.S. Characteristics of honey bee (*Apis mellifera carnica*, Pollman 1879) queens reared in Slovenian commercial breeding stations. *J. Apic. Sci.* **2015**, *59*, 5–12. [CrossRef]
19. Uzunov, A.; Brascamp, E.W.; Büchler, R. The basic concept of honey bee breeding programs. *Bee World* **2017**, *94*, 84–87. [CrossRef]
20. Büchler, R.; Andonov, S.; Bernstein, R.; Bienefeld, K.; Costa, C.; Du, M.; Gabel, M.; Given, K.; Hatjina, F.; Harpur, B.A.; et al. Standard methods for rearing and selection of *Apis mellifera* queens 2.0. *J. Apic. Res.* **2024**, *63*, 1–57. [CrossRef]
21. Büchler, R.; Berg, S.; Le Conte, Y. Breeding for resistance to *Varroa destructor* in Europe. *Apidologie* **2010**, *41*, 393–408. [CrossRef]
22. Rinderer, T.E.; Harris, J.W.; Hunt, G.J.; De Guzman, L.I. Breeding for resistance to *Varroa destructor* in North America. *Apidologie* **2010**, *41*, 409–424. [CrossRef]
23. van Alphen, J.J.; Fernhout, B.J. Natural selection, selective breeding, and the evolution of resistance of honeybees (*Apis mellifera*) against Varroa. *Zool. Lett.* **2020**, *6*, 6. [CrossRef] [PubMed]
24. Gregorc, A.; Bakonyi, T. Viral infections in queen bees (*Apis mellifera carnica*) from rearing apiaries. *Acta Vet. Brno* **2012**, *81*, 15–19. [CrossRef]
25. Žvokelj, L.; Bakonyi, T.; Korošec, T.; Gregorc, A. Appearance of acute bee paralysis virus, black queen cell virus and deformed wing virus in Carnolian honey bee (*Apis mellifera carnica*) queen rearing. *J. Apic. Res.* **2020**, *59*, 53–58. [CrossRef]
26. Domingues, C.E.; Šimenc, L.; Toplak, I.; de Graaf, D.C.; De Smet, L.; Verbeke, W.; Peelman, L.; Annaloni, L.S., Gregorc, A. Eggs sampling as an effective tool for identifying the incidence of viruses in honey bees involved in artificial queen rearing. *Sci. Rep.* **2024**, *14*, 9612. [CrossRef]
27. Gray, A.; Brodschneider, R.; Adjlane, N.; Ballis, A.; Brusbardis, V.; Charrière, J.-D.; Chlebo, R.; Coffey, M.F.; Cornelissen, B.; Costa, C.A.; et al. Loss rates of honey bee colonies during winter 2017/18 in 36 countries participating in the COLOSS survey, including effects of forage sources. *J. Apic. Res.* **2019**, *58*, 479–485. [CrossRef]
28. Goulson, D.; Nicholls, E.; Botías, C.; Rotheray, E.L. Bee declines driven by combined stress from parasites, pesticides, and lack of flowers. *Science* **2015**, *347*, 1255957. [CrossRef]
29. Neov, B.; Georgieva, A.; Shumkova, R.; Radoslavov, G.; Hristov, P. Biotic and abiotic factors associated with colonies mortalities of managed honey bee (*Apis mellifera*). *Diversity* **2019**, *11*, 237. [CrossRef]
30. Smart, M.D.; Otto, C.R.; Lundgren, J.G. Nutritional status of honey bee (*Apis mellifera* L.) workers across an agricultural land-use gradient. *Sci. Rep.* **2019**, *9*, 16252. [CrossRef]
31. Winkler, K.; Fuchs, R.; Rounsevell, M.; Herold, M. Global land use changes are four times greater than previously estimated. *Nat. Commun.* **2021**, *12*, 2501. [CrossRef] [PubMed]
32. de Groot, G.S.; Aizen, M.A.; Sáez, A.; Morales, C.L. Large-scale monoculture reduces honey yield: The case of soybean expansion in Argentina. *Agric. Ecosyst. Environ.* **2021**, *306*, 107203. [CrossRef]
33. Di Pasquale, G.; Salignon, M.; Le Conte, Y.; Belzunces, L.P.; Decourtye, A.; Kretzschmar, A.; Suchail, S.; Brunet, J.-L.; Alaux, C. Influence of pollen nutrition on honey bee health: Do pollen quality and diversity matter? *PLoS ONE* **2013**, *8*, e72016. [CrossRef] [PubMed]
34. Brodschneider, R.; Crailsheim, K. Nutrition and health in honey bees. *Apidologie* **2010**, *41*, 278–294. [CrossRef]
35. Lamontagne-Drolet, M.; Samson-Robert, O.; Giovenazzo, P.; Fournier, V. The impacts of two protein supplements on commercial honey bee (*Apis mellifera* L.) colonies. *J. Apic. Res.* **2019**, *58*, 800–813. [CrossRef]
36. Paiva, J.P.L.M.; Esposito, E.; de Morais Honorato De Souza, G.I.; Francoy, T.M.; Morais, M.M. Effects of ensiling on the quality of protein supplements for honey bees *Apis mellifera*. *Apidologie* **2019**, *50*, 414–424. [CrossRef]
37. Peirson, M.; Ibrahim, A.; Ovinge, L.P.; Hoover, S.E.; Guarna, M.M.; Melathopoulos, A.; Pernal, S.F. The effects of protein supplementation, fumagillin treatment, and colony management on the productivity and long-term survival of honey bee (*Apis mellifera*) colonies. *PLoS ONE* **2024**, *19*, e0288953. [CrossRef]
38. Hrassnigg, N.; Crailsheim, K. Differences in drone and worker physiology in honeybees (*Apis mellifera* L.). *Apidologie* **2005**, *36*, 255–277. [CrossRef]

39. vanEngelsdorp, D.; Meixner, M.D. A historical review of managed honey bee populations in Europe and the United States and the factors that may affect them. *J. Invertebr. Pathol.* **2010**, *103*, S80–S95. [CrossRef]
40. Menzel, R. The honeybee as a model for understanding the basis of cognition. *Nat. Rev. Neurosci.* **2012**, *13*, 758–768. [CrossRef]
41. Tsuruda, J.M.; Chakrabarti, P.; Sagili, R.R. Honey bee nutrition. *Vet. Clin. N. Am. Food Anim. Pract.* **2021**, *37*, 505–519. [CrossRef] [PubMed]
42. Liolios, V.; Tananaki, C.; Dimou, M.; Kanelis, D.; Goras, G.; Karazafiris, E.; Thrasyvoulou, A. Ranking pollen from bee plants according to their protein contribution to honey bees. *J. Apic. Res.* **2015**, *54*, 582–592. [CrossRef]
43. Radev, Z. Variety in protein content of pollen from 50 plants from Bulgaria. *Bee World* **2018**, *95*, 81–83. Available online: https://www.tandfonline.com/doi/full/10.1080/0005772X.2018.1486276#d1e88 (accessed on 28 November 2024). [CrossRef]
44. Roulston, T.A.H.; Cane, J.H.; Buchmann, S.L. What governs protein content of pollen: Pollinator preferences, pollen–pistil interactions, or phylogeny? *Ecol. Monogr.* **2000**, *70*, 617–643. [CrossRef]
45. Dolezal, A.G.; Carrillo-Tripp, J.; Judd, T.M.; Allen Miller, W.; Bonning, B.C.; Toth, A.L. Interacting stressors matter: Diet quality and virus infection in honeybee health. *R. Soc. Open Sci.* **2019**, *6*, 181803. [CrossRef] [PubMed]
46. Bryś, M.S.; Skowronek, P.; Strachecka, A. Pollen Diet—Properties and Impact on a Bee Colony. *Insects* **2021**, *12*, 798. [CrossRef] [PubMed]
47. Fewell, J.H.; Winston, M.L. Colony state and regulation of pollen foraging in the honey bee, *Apis mellifera* L. *Behav. Ecol. Sociobiol.* **1992**, *30*, 387–393. [CrossRef]
48. Danner, N.; Keller, A.; Härtel, S.; Steffan-Dewenter, I. Honey bee foraging ecology: Season but not landscape diversity shapes the amount and diversity of collected pollen. *PLoS ONE* **2017**, *12*, e0183716. [CrossRef]
49. Wille, H.; Wille, M.; Kilchenmann, V.; Imdorf, A.; Bühlmann, G. Pollenernte und Massenwechsel von drei *Apis mellifera*-Völkern auf demselben Bienenstand in zwei aufeinan-derfolgenden Jahren. *Rev. Suiss Zool.* **1985**, *92*, 897–914. [CrossRef]
50. Avni, D.; Hendriksma, H.P.; Dag, A.; Uni, Z.; Shafir, S. Nutritional aspects of honey bee-collected pollen and constraints on colony development in the eastern Mediterranean. *J. Insect Physiol.* **2014**, *69*, 65–73. [CrossRef]
51. Rortais, A.; Arnold, G.; Halm, M.-P.; Touffet-Briens, F. Modes of honeybees exposure to systemic insecticides: Estimated amounts of contaminated pollen and nectar consumed by different categories of bees. *Apidologie* **2005**, *36*, 71–83. [CrossRef]
52. Wright, G.A.; Nicolson, S.W.; Shafir, S. Nutritional physiology and ecology of honey bees. *Annu. Rev. Entomol.* **2018**, *63*, 327–344. [CrossRef] [PubMed]
53. Radev, Z.; Liolios, V.; Tananaki, C.; Thrasyvoulou, A. The impact of the nutritive value of pollen on the development, reproduction and productivity of honey bee (*Apis mellifera* L.). *Bulg. J. Agric. Sci.* **2014**, *20*, 685–689.
54. Radev, Z. The impact of different protein content of pollen on honey bee (*Apis mellifera* L.) reproduction. *New J. Sci* **2019**, *8*, 80–88.
55. Frias, B.E.D.; Barbosa, C.D.; Laurenco, A.P. Pollen nutrition in honey bees (*Apis mellifera*): Impact on adult health. *Apidologie* **2016**, *47*, 15–25. [CrossRef]
56. Crailsheim, K.; Schneider, L.H.W.; Hrassnigg, N.; Bühlmann, G.; Brosch, U.; Gmeinbauer, R.; Schöffmann, B. Pollen consumption and utilization in worker honeybees (*Apis mellifera carnica*): Dependence on individual age and function. *J. Insect Physiol.* **1992**, *38*, 409–419. [CrossRef]
57. Li, C.; Xu, B.; Wang, Y.; Yang, Z.; Yang, W. Protein content in larval diet affects adult longevity and antioxidant gene expression in honey bee workers. *Entomol. Exp. Appl.* **2014**, *151*, 19–26. [CrossRef]
58. Schmehl, D.R.; Teal, P.E.; Frazier, J.L.; Grozinger, C.M. Genomic analysis of the interaction between pesticide exposure and nutrition in honey bees (*Apis mellifera*). *J. Insect Physiol.* **2014**, *71*, 177–190. [CrossRef]
59. Kieliszek, M.; Piwowarek, K.; Kot, A.M.; Błażejak, S.; Chlebowska-Śmigiel, A.; Wolska, I. Pollen and bee bread as new health-oriented products: A review. *Trends Food Sci. Technol.* **2018**, *71*, 170–180. [CrossRef]
60. Bakour, M.; Fernandes, Â.; Barros, L.; Sokovic, M.; Ferreira, I.C. Bee bread as a functional product: Chemical com-position and bioactive properties. *LWT* **2019**, *109*, 276–282. [CrossRef]
61. Gregorc, A.; Sampson, B.; Knight, O.R.; Adamczyk, J. Diet quality affects honey bee (Hymenoptera: Apidae) mortality under laboratory conditions. *J. Apic. Res.* **2019**, *58*, 492–493. [CrossRef]
62. DeGrandi-Hoffman, G.; Wardell, G.; Ahumada-Segura, F.; Rinderer, T.; Danka, R.; Pettis, J. Comparisons of pollen substitute diets for honey bees: Consumption rates by colonies and effects on brood and adult populations. *J. Apic. Res.* **2008**, *47*, 265–270. [CrossRef]
63. Zheng, B.; Wu, Z.; Xu, B. The effects of dietary protein levels on the population growth, performance, and physiology of honey bee workers during early spring. *J. Insect Sci.* **2014**, *14*, 191. [CrossRef] [PubMed]
64. García-Vicente, E.J.; Martín, M.; Rey-Casero, I.; Pérez, A.; Martín, J.; García, A.; Alonso, J.M.; Risco, D. Effects of feeding with a protein liquid supplement on productivity, mortality and health of *Apis mellifera* hives in southwestern Spain. *Res. J. Vet. Sci.* **2024**, *169*, 105173. [CrossRef]

65. DeGrandi-Hoffman, G.; Chen, Y.; Rivera, R.; Carroll, M.; Chambers, M.; Hidalgo, G.; de Jong, E.W. Honey bee colonies provided with natural forage have lower pathogen loads and higher overwinter survival than those fed protein supplements. *Apidologie* **2016**, *47*, 186–196. [CrossRef]
66. Branchiccela, B.; Castelli, L.; Corona, M.; Díaz-Cetti, S.; Invernizzi, C.; Martínez De La Escalera, G.; Mendoza, Y.; Santos, E.; Silva, C.; Zunino, P.; et al. Impact of nutritional stress on the honeybee colony health. *Sci. Rep.* **2019**, *9*, 10156. [CrossRef]
67. Giacobino, A.; Pacini, A.; Molineri, A.; Bulacio-Cagnolo, N.; Merke, J.; Orellano, E.; Gaggiotii, M.; Signorini, M. Impact of nutritional and sanitary management on *Apis mellifera* colony dynamics and pathogen loads. *SJAR* **2022**, *20*, e0305. [CrossRef]
68. Haydak, M.H. Honey bee nutrition. *Annu. Rev. Entomol.* **1970**, *15*, 143–156. [CrossRef]
69. Nicolson, S.W. Sweet solutions: Nectar chemistry and quality. *Philos. Trans. R. Soc. B* **2022**, *377*, 20210163. [CrossRef]
70. Venjakob, C.; Ruedenauer, F.A.; Klein, A.M.; Leonhardt, S.D. Variation in nectar quality across 34 grassland plant species. *Plant Biol.* **2022**, *24*, 134–144. [CrossRef]
71. Ali, J.; Abbas, A.; Abbas, S.; Ji, Y.; Khan, K.A.; Ghramh, H.A.; Mahamood, M.; Chen, R. Honeydew: A keystone in insect–plant interactions, current insights and future perspectives. *J. Appl. Entomol.* **2024**, *148*, 727–733. [CrossRef]
72. Kunieda, T.; Fujiyuki, T.; Kucharski, R.; Foret, S.; Ament, S.A.; Toth, A.L.; Ohashi, K.; Takeuchi, H.; Kamikouchi, A.; Kage, E.; et al. Carbohydrate metabolism genes and pathways in insects: Insights from the honey bee genome. *Insect Mol. Biol.* **2006**, *15*, 563–576. [CrossRef] [PubMed]
73. Pamminger, T.; Becker, R.; Himmelreich, S.; Schneider, C.W.; Bergtold, M. The nectar report: Quantitative review of nectar sugar concentrations offered by bee visited flowers in agricultural and non-agricultural landscapes. *PeerJ* **2019**, *7*, e6329. [CrossRef] [PubMed]
74. Kim, W.; Gilet, T.; Bush, J.W. Optimal concentrations in nectar feeding. *Proc. Natl. Acad. Sci. USA* **2011**, *108*, 16618–16621. [CrossRef]
75. Sammataro, D.; Weiss, M. Comparison of productivity of colonies of honey bees, *Apis mellifera*, supplemented with sucrose or high fructose corn syrup. *J. Insect Sci.* **2013**, *13*, 19. [CrossRef]
76. Adler, L.S. The ecological significance of toxic nectar. *Oikos* **2000**, *91*, 409–420. [CrossRef]
77. Clinch, P.G.; Palmer-Jones, T.; Forster, I.W. Effect on honey bees of nectar from the yellow kowhai (*Sophora microphylla* Ait.). *N. Zeal. J. Agr. Res.* **1972**, *15*, 194–201. [CrossRef]
78. Paula, V.F.; Barbosa, L.C.A.; Demuner, A.J.; Campos, L.A.; Pinheiro, A.L. Entomotoxicity of the nectar from *Ochroma lagopus* Swartz (Bombacaceae). *Cien. Cult.* **1997**, *49*, 274–277.
79. Reinhard, A.; Janke, M.; von der Ohe, W.; Kempf, M.; Theuring, C.; Hartmann, T.; Schreier, P.; Beuerle, T. Feeding deterrence and detrimental effects of pyrrolizidine alkaloids fed to honey bees (*Apis mellifera*). *J. Chem. Ecol.* **2009**, *35*, 1086–1095. [CrossRef]
80. Lucchetti, M.A.; Kilchenmann, V.; Glauser, G.; Praz, C.; Kast, C. Nursing protect honey bee larvae from secondary metabolites of pollen. *Proc. R. Soc. B* **2018**, *285*, 20172849. [CrossRef]
81. Brugnerotto, P.; Seraglio, S.K.T.; Schulz, M.; Gonzaga, L.V.; Fett, R.; Costa, A.C.O. Pyrrolizidine alkaloids and beehive products: A review. *Food Chem.* **2021**, *342*, 128384. [CrossRef] [PubMed]
82. Alvarado-Avila, L.Y.; Moguel-Ordóñez, Y.B.; García-Figueroa, C.; Ramírez-Ramírez, F.J.; Arechavaleta-Velasco, M.E. Presence of pyrrolizidine alkaloids in honey and the effects of their consumption on humans and honeybees. *Review. Rev. Mex. Cienc. Pecu.* **2022**, *13*, 787–802. [CrossRef]
83. EFSA Panel on Contaminants in the Food Chain (CONTAM). Scientific opinion on pyrrolizidine alkaloids in food and feed. *EFSA J.* **2011**, *9*, 2406. [CrossRef]
84. EFSA (European Food Safety Authority). Dietary exposure assessment to pyrrolizidine alkaloids in the European population. *EFSA J.* **2016**, *14*, 4572. [CrossRef]
85. EFSA Panel on Contaminants in the Food Chain (CONTAM). Risks for human health related to the presence of pyrrolizidine alkaloids in honey, tea, herbal infusions and food supplements. *EFSA J.* **2017**, *15*, 4908. [CrossRef]
86. Paoli, P.P.; Donley, D.; Stabler, D.; Saseendranath, A.; Nicolson, S.W.; Simpson, S.J.; Wright, G.A. Nutritional balance of essential amino acids and carbohydrates of the adult worker honeybee depends on age. *Amino Acids* **2014**, *46*, 1449–1458. [CrossRef]
87. Couvillon, M.J.; Riddell Pearce, F.C.; Accleton, C.; Fensome, K.A.; Quah, S.K.; Taylor, E.L.; Ratnieks, F.L. Honey bee foraging distance depends on month and forage type. *Apidologie* **2015**, *46*, 61–70. [CrossRef]
88. Silva, P.M.; Gauche, C.; Gonzaga, L.V.; Costa, A.C.O.; Fett, R. Honey: Chemical composition, stability and authenticity. *Food Chem.* **2016**, *196*, 309–323. [CrossRef]
89. Doner, L.W. The sugars of honey—A review. *JSFA* **1977**, *28*, 443–456. [CrossRef]
90. Wheeler, M.; Robinson, G. Diet-dependent gene expression in honey bees: Honey vs. sucrose or high fructose corn syrup. *Sci. Rep.* **2014**, *4*, 5726. [CrossRef]
91. Mogren, C.L.; Margotta, J.; Danka, R.G.; Healy, K. Supplemental carbohydrates influence abiotic stress resistance in honey bees. *J. Apic. Res.* **2018**, *57*, 682–689. [CrossRef]

92. Quinlan, G.; Döke, M.A.; Ortiz-Alvarado, Y.; Rodriguez-Gomez, N.; Koru, Y.B.; Underwood, R. Carbohydrate nutrition associated with health of overwintering honey bees. *J. Insect Sci.* **2023**, *23*, 16. [CrossRef] [PubMed]
93. Frizzera, D.; Del Fabbro, S.; Ortis, G.; Zanni, V.; Bortolomeazzi, R.; Nazzi, F.; Annoscia, D. Possible side effects of sugar supplementary nutrition on honey bee health. *Apidologie* **2020**, *51*, 594–608. [CrossRef]
94. Dobson, H.E. Survey of pollen and pollenkitt lipids—Chemical cues to flower visitors? *Am. J. Bot.* **1988**, *75*, 170–182. [CrossRef]
95. Ischebeck, T. Lipids in pollen—They are different. *BBA-Mol. Cell. Biol. Lipids* **2016**, *1861*, 1315–1328. [CrossRef]
96. Arrese, E.L.; Soulages, J.L. Insect fat body: Energy, metabolism, and regulation. *Annu. Rev. Entomol.* **2010**, *55*, 207–225. [CrossRef]
97. Cantrill, R.C.; Hepburn, H.R.; Warner, S.J. Changes in lipid composition during sealed brood development of African worker honeybees. *CBP* **1981**, *68*, 351–353. [CrossRef]
98. Skowronek, P.; Wójcik, Ł.; Strachecka, A. Fat body—Multifunctional insect tissue. *Insects* **2021**, *12*, 547. [CrossRef]
99. Corby-Harris, V.; Snyder, L.; Meador, C. Fat body lipolysis connects poor nutrition to hypopharyngeal gland degradation in *Apis mellifera*. *J. Insect Physiol.* **2019**, *116*, 1–9. [CrossRef]
100. Stabler, D.; Al-Esawy, M.; Chennells, J.A.; Perri, G.; Robinson, A.; Wright, G.A. Regulation of dietary intake of protein and lipid by nurse-age adult worker honeybees. *J. Exp. Biol.* **2021**, *224*, jeb230615. [CrossRef]
101. Elizabeth Deeter, M.; Snyder, L.A.; Meador, C.; Corby-Harris, V. Accelerated abdominal lipid depletion from pesticide treatment alters honey bee pollen foraging strategy, but not onset, in worker honey bees. *J. Exp. Biol.* **2023**, *226*, jeb245404. [CrossRef] [PubMed]
102. Manning, R. Fatty acids in pollen: A review of their importance for honey bees. *Bee World* **2001**, *82*, 60–75. [CrossRef]
103. Robinson, F.A.; Nation, J.L. Long-chain fatty acids in honeybees in relation to sex, caste, and food during development. *J. Apic. Res.* **1970**, *9*, 121–127. [CrossRef]
104. Herbert, E.W., Jr.; Shimanuki, H.; Shasha, B.S. Brood rearing and food consumption by honeybee colonies fed pollen substitutes supplemented with starch-encapsulated pollen extracts. *J. Apic. Res.* **1980**, *19*, 115–118. [CrossRef]
105. Arien, Y.; Dag, A.; Yona, S.; Tietel, Z.; Cohen, T.L.; Shafir, S. Effect of diet lipids and omega-6: 3 ratio on honey bee brood development, adult survival and body composition. *J. Insect Physiol.* **2020**, *124*, 104074. [CrossRef]
106. Negri, P.; Ramirez, L.; Quintana, S.; Szawarski, N.; Maggi, M.; Conte, Y.L.; Lamattina, L.; Eguaras, M. Dietary supplementation of honey bee larvae with arginine and abscisic acid enhances nitric oxide and granulocyte immune responses after trauma. *Insects* **2017**, *8*, 85. [CrossRef]
107. Inouye, D.W.; Waller, G.D. Responses of honey bees (*Apis mellifera*) to amino acid solutions mimicking floral nectars. *Ecology* **1984**, *65*, 618–625. [CrossRef]
108. Locatelli, F.; Bundrock, G.; Müller, U. Focal and temporal release of glutamate in the mushroom bodies improves olfactory memory in *Apis mellifera*. *J. Neurosci.* **2005**, *25*, 11614–11618. [CrossRef]
109. Bryś, M.S.; Strachecka, A. The key role of amino acids in pollen quality and honey bee physiology—A review. *Molecules* **2024**, *9*, 2605. [CrossRef]
110. Somerville, D.C.; Nicol, H.I. Crude protein and amino acid composition of honey bee collected pollen pellets from south-east Australia and a note on laboratory disparity. *Aust. J. Exp. Agric.* **2006**, *46*, 141–149. [CrossRef]
111. De Groot, A.P. Amino acid requirements for growth of the honeybee (*Apis mellifica* L.). *Experientia* **1952**, *8*, 192–194. [CrossRef] [PubMed]
112. Elsayeh, W.A.; Cook, C.; Wright, G.A. B-vitamins influence the consumption of macronutrients in honey bees. *Front. Sustain. Food Syst.* **2022**, *6*, 804002. [CrossRef]
113. Hassanyar, A.K.; Huang, J.; Nie, H.; Li, Z.; Hussain, M.; Rizwan, M.; Su, S. The association between the hypopharyngeal glands and the molecular mechanism which honey bees secrete royal jelly. *J. Apic. Res.* **2023**, 1–16. [CrossRef]
114. Collazo, N.; Carpena, M.; Nuñez-Estevez, B.; Otero, P.; Simal-Gandara, J.; Prieto, M.A. Health promoting properties of bee royal jelly: Food of the queens. *Nutrients* **2021**, *13*, 543. [CrossRef]
115. Howe, S.R.; Dimick, P.S.; Benton, A.W. Composition of freshly harvested and commercial royal jelly. *J. Apic. Res.* **1985**, *24*, 52–61. [CrossRef]
116. Liming, W.; Jinhui, Z.; Xiaofeng, X.; Yi, L.; Jing, Z. Fast determination of 26 amino acids and their content changes in royal jelly during storage using ultra-performance liquid chromatography. *J. Food Compos. Anal.* **2009**, *22*, 242–249. [CrossRef]
117. Farjan, M.; Dmitryjuk, M.; Lipiński, Z.; Biernat-Łopieńska, E.; Żółtowska, K. Supplementation of the honey bee diet with vitamin C: The effect on the antioxidative system of *Apis mellifera carnica* brood at different stages. *J. Apic. Res.* **2012**, *51*, 263–270. [CrossRef]
118. Glavinic, U.; Stankovic, B.; Draskovic, V.; Stevanovic, J.; Petrovic, T.; Lakic, N.; Stanimirovic, Z. Dietary amino acid and vitamin complex protects honey bee from immunosuppression caused by *Nosema ceranae*. *PLoS ONE* **2017**, *12*, e0187726. [CrossRef]
119. Stanimirović, Z.; Glavinić, U.; Ristanić, M.; Jelisić, S.; Vejnović, B.; Niketić, M.; Stevanović, J. Diet supplementation helps honey bee colonies in combat infections by enhancing their hygienic behaviour. *Acta Vet.* **2022**, *72*, 145–166. [CrossRef]
120. Nicolson, S.W. Water homeostasis in bees, with the emphasis on sociality. *J. Exp. Biol.* **2009**, *212*, 429–434. [CrossRef]

121. Filipiak, M.; Kuszewska, K.; Asselman, M.; Denisow, B.; Stawiarz, E.; Woyciechowski, M.; Weiner, J. Ecological stoichiometry of the honeybee: Pollen diversity and adequate species composition are needed to mitigate limitations imposed on the growth and development of bees by pollen quality. *PLoS ONE* **2017**, *12*, e0183236. [CrossRef] [PubMed]
122. Herbert, J.; Elton, W.; Shimanuki, H.S. Mineral requirements for brood-rearing by honeybees fed a synthetic diet. *J. Apic. Res.* **1978**, *17*, 118–122. [CrossRef]
123. Lau, P.W.; Nieh, J.C. Salt preferences of honey bee water foragers. *J. Exp. Biol.* **2016**, *219*, 790–796. [CrossRef] [PubMed]
124. Khan, K.A.; Ghramh, H.A.; Ahmad, Z.; El-Niweiri, M.A.; Mohammed, M.E.A. Honey bee (*Apis mellifera*) preference towards micronutrients and their impact on bee colonies. *Saudi J. Biol. Sci.* **2021**, *28*, 3362–3366. [CrossRef]
125. Cairns, S.M.; Wratten, S.D.; Filipiak, M.; Veronesi, E.R.; Saville, D.J.; Shields, M.W. Ratios rather than concentrations of nutritionally important elements may shape honey bee preferences for 'dirty water'. *Ecol. Entomol.* **2021**, *46*, 1236–1240. [CrossRef]

insects

MDPI

Article

How Science Supports Honey Bees: Identification of Research on Best Practices in Beekeeping

Kristina Gratzer *, Veronika Musalkova and Robert Brodschneider

Department of Biology, University of Graz, 8010 Graz, Austria; robert.brodschneider@uni-graz.at (R.B.)
* Correspondence: kristina.gratzer@uni-graz.at

Simple Summary: Honey bees are essential for food production and healthy ecosystems, but they face many threats from pests, diseases, and poor management. Beekeepers around the world use different practices to keep their colonies healthy, yet it is often unclear which methods work best. In this study, we collected and analyzed results from 191 scientific field studies published since 1995. We focused only on research that tested practices directly on honey bee colonies under real beekeeping conditions. Altogether, we found 744 different practice records and evaluated their effects on colony health and productivity. Most studies dealt with controlling the *Varroa* mite, one of the most damaging pests of honey bees. Other important topics were the management of colonies, the prevention of american foulbrood disease, and the control of other parasites and pests. We also looked at when during the year the practices were applied and in which regions of the world. The results show that there is no single solution, but a variety of methods can help improve colony health. This study provides a practical framework that can guide beekeepers, advisors, and policymakers in making evidence-based decisions to support sustainable apiculture for beekeepers, veterinarians, and extension professionals.

Abstract: Honey bee health and productivity are strongly linked to management practices and biosecurity measures. We collected and analyzed 744 practice records from 191 peer-reviewed field studies published since 1995, each documenting the impact of a specific hive intervention on colony health or productivity parameters. Practices were categorized into good beekeeping practices (n = 128, 17.2%) and biosecurity measures (n = 616, 82.8%) and grouped by management or pathogen theme, geographic region, and season. Most research originated from Europe (34.6%) and North America (33.4%), with nearly all focused on *Apis mellifera* (99.9%). *Varroa* control dominated (57.0%), followed by general apiary management (17.2%) and american foulbrood (9.7%). For varroosis, "soft" acaricides such as oxalic and formic acid accounted for 58.5% of records, while "hard" synthetic products represented 21.0%. Within general apiary management, colony management was most frequent (46.9%). For american foulbrood, antibiotics (41.7%) and biotechnical methods (22.2%) were prominent. *Tropilaelaps* app. treatments relied mainly on "soft" acaricides (81.5%), while small hive beetle control focused on in-hive traps (55.9%). Seasonally, most interventions occurred from August to October, with AFB measures peaking in spring and early summer. The dataset highlights regional research gaps and offers a structured, expandable framework to guide future research and support evidence-based decision-making in beekeeping and advisory services.

Keywords: hive management; *Apis mellifera*; *Aethina tumida*; *Varroa destructor*; *Paenibacillus larvae*

1. Introduction

Health and productivity of livestock depend on management practices and biosecurity measures. Good beekeeping management keeps livestock protected from major stressors, reduces the risk of diseases, and positively influences the productivity and economic profit of the farmer [1,2]. Managed honey bees as livestock are housed in artificial structures and subject to human selection, yet they forage freely in the landscape. According to the World Organisation for Animal Health (OIE), bees are classified as terrestrial animals and thus fall under veterinary care [3]. Honey bees are the most economically important pollinators globally, significantly contributing to agricultural productivity and ecosystem stability [4]. In addition, bees provide a wide variety of products: honey, pollen, propolis, wax, royal jelly, and venom. Beekeepers also sell colonies and rear queens, further diversifying their economic activities [1].

Large-scale honey bee colony losses have been reported worldwide. Numerous surveys and research projects have sought to identify the primary drivers of these losses [5–9]. These studies often highlight the *Varroa* mite (*Varroa destructor*) as the major threat to *A. mellifera* colonies. Originating on *Apis cerana*, *V. destructor* exploits the lack of a natural host–parasite balance, leading to colony collapse within a couple of years if untreated. This parasite compromises bee health by weakening immune responses [10,11], transmitting viruses [12,13], and impairing physiological development [14,15].

Bacterial infections remain a major contributor to honey bee colony losses worldwide. Among them, american Foulbrood (AFB), caused by the spore-forming *Paenibacillus larvae*, and european foulbrood (EFB), caused by *Melissococcus plutonius*, are particularly relevant due to their persistence and ease of transmission [16,17]. Further, EFB infections are often complemented by secondary pathogens such as *Paenibacillus alvei*, *Brevibacillus laterosporus*, and *Enterococcus faecalis* [16]. Beyond bacterial pathogens, parasitic mites also pose threats to honey bees. *Tropilaelaps* mites (*Tropilaelaps* spp.), originally parasitic on *Apis dorsata*, represent another significant threat, particularly in Asia. They outcompete *Varroa* mites under certain conditions due to their shorter phoretic stage and rapid reproduction [18–20]. Infestations can lead to brood mortality and colony decline [21]. Continuous monitoring is therefore essential, as highlighted by ref. [22], who introduced the "rapid brood decapping" method to assess *T. mercedesae* infestation in brood. While this work does not provide direct evidence of westward spread into Europe, it underscores the importance of surveillance tools to track potential range expansions. The small hive beetle (*Aethina tumida*), an invasive species originating from Africa, has also established itself in several parts of the world, including the Americas, Australia, and more recently in Europe [23,24]. It damages colonies by consuming brood, pollen, and honey, and its presence can ultimately trigger colony collapse [25].

Such biological threats gain further relevance in the context of globally practiced beekeeping and the diverse environmental conditions in which honey bees are now kept. Originally limited to the Old World, honey bees have since been introduced into a wide range of climates and landscapes by humans. The primary species in apiculture is the Western honey bee *A. mellifera*, followed by the Eastern honey bee *A. cerana*; they have adapted to diverse environmental conditions, demanding region-specific beekeeping practices [26,27]. Comprehensive measures for prevention, early detection, treatment, and pest control are essential for protecting colony health and ensuring the long-term sustainability of apiculture [3].

This study aims to identify beekeeping and biosecurity practices with demonstrated effects on colony health and productivity. To achieve this, we analyzed peer-reviewed field studies that included comparisons between treatment and control groups under realistic on-field conditions. Practices were categorized into good beekeeping practices, referring to regular hive management actions such as timely feeding, queen replacement, or hygienic comb renewal, and biosecurity measures, which focus on preventing the introduction and spread of pathogens within and between apiaries, including quarantine protocols or disinfection routines [1,3]. The impact of each practice on colony health and productivity was evaluated to identify practical recommendations for beekeepers, veterinarians, and advisory services. Results are presented in both tabular and graphical form and are contextualized by region and season.

Although numerous studies on honey bee health exist [28–30], comprehensive syntheses that translate these findings into actionable, field-tested management recommendations for stakeholders remain limited. Reasons for this include inconsistent study outcomes, differences in regulations, and region-specific applicability. For example, the path from pathogen identification in the laboratory to the availability of effective treatment options for beekeepers is often long and uncertain. This work is intended as a framework for the structured analysis of applied honey bee research and is designed to be flexible enough for future expansion as new studies become available. The categorization is based on the lists of practices compiled and prioritized by [1,3].

2. Materials and Methods

To ensure practical relevance and comparability, we focused on studies in which the subject unit was a honey bee colony, the research was conducted under field conditions, and both experimental and control groups were included. Only publications from 1995 onwards in English language were considered. By applying these criteria, we established a structured framework that synthesizes existing applied research and is designed to be continuously expandable as new studies become available.

All scientific publications were obtained through the Google Scholar search engine. Out of 341 reviewed papers, 191 were included in this study, considering a total of 12,400 beehives used for their field studies. Papers representing reviews or survey data were excluded; however, these were sometimes useful for obtaining an overview of the research conducted so far on the related theme. Further excluding criteria cover papers based on laboratory experiments or studies where a single honey bee or a group of bees in cages represented the studied object.

Our next step was to compile all relevant data into a MS Excel spreadsheet (Microsoft Corporation, Redmond, WA, USA), which served as the basis for the summary analyses. This spreadsheet, referred to as Supplemental Material S1, represents the core output of this work. Each reviewed publication was systematically screened for relevant practices. In the resulting table, each row corresponds to a single practice record, defined as one documented application of a specific practice and its observed effect. A single publication may contain multiple practices, or a single practice evaluated for multiple outcomes, each of which is represented as a separate practice record. For example, caging the queen may reduce *Varroa* mite infestation levels (one row), while also affecting honey yields (a second row). Throughout this study, we therefore use the term practice record to describe these individual entries. To ensure clarity and consistency in terminology, practice refers to a specific beekeeping measure or management approach investigated in a study, while practice record denotes its application and evaluation in a particular experimental context. The term sub-theme is used to describe categories within broader thematic groups (e.g., "soft acaricides" within the theme varroosis). When referring specifically to interventions against pests or pathogens, particularly in the context of varroosis, *Tropilaelaps* spp., or american foulbrood, we use the term treatment.

To allow for comparability across studies with varying methodologies, each practice record was described as precisely as possible while maintaining a simplified, standardized format. For example, numerous methods have been investigated in research in an effort to combat varroosis. Additional technical details such as mode of application or substance concentration were included in separate columns where available, whereas more specific information on the practice record must be retrieved from the original paper. As the practices were identified, these were subsequently categorized into eight thematic areas: general apiary management, small hive beetle infestation (*Aethina tumida*), american foulbrood (AFB), chalkbrood, european foulbrood (EFB), greater wax moth (*Galleria mellonella*), *Tropilaelaps* spp. infestation, and *Varroa* mite infestation [1,3] (Supplemental Material S1). For the sub-themes AFB and varroosis, we conducted additional analyses. Specifically, we examined the seasonality of practice application for both sub-themes. For varroosis, we also assessed the efficacy of practices, with a particular focus on the most frequently studied practice type: treatments involving oxalic acid.

For the category of "biosecurity measures", the impact was further categorized to provide a more precise classification of significant efficacy. Positive effects were subdivided into three levels: "positive" for efficacy greater than 85%, "moderate positive" for efficacy between 50% and 85%, and "low positive" for efficacy below 50%. Measures with less than 10% efficacy were classified as having "no effect". Efficacy data, means and numbers of application, concentration, length of the procedure and dosage are provided in separate columns. If unavailable, the classification from the research paper is used. Unless otherwise stated, no additional statistical analysis was conducted, and effect sizes or significance levels were not recalculated. Therefore, the frequency values presented in this work reflect how often certain practices were associated with specific outcomes in the reviewed studies, rather than indicating the strength or statistical significance of those effects. As such, results should be interpreted as descriptive evidence of reported associations, not as quantified effect estimates.

For the geographic classification we used a simplified subdivision ("continent", "country", "region") with the main emphasis put on the division of European regions. "Continent" includes Europe, North America, Central America, South America, Asia, Africa, Oceania; "region" is only used for European and includes division into six regions: Northern, Western, Central, Southern, Eastern and South-eastern Europe.

Regarding the seasonal data, we recorded these in the Supplemental Material S1 by month. To ensure consistency between hemispheres, seasonal data from the Southern Hemisphere were adjusted by shifting six months forward, aligning July–December with January–June of the Northern Hemisphere. The column "species" relates to the biological classification of the investigated honey bee colonies.

Additionally, we also recorded the year of publication and year of the experiment, number of colonies involved in the research and the reference as well. When necessary, we also added a comment in a separate column. For visualization, the software GraphPad Prism 8.0.2. (GraphPad Software, San Diego, CA, USA) was used.

3. Results

3.1. Geographic and Temporal Distribution

Most research in our study originates from Europe (34.6%) and North America (33.4%), followed by Asia (19.5%), Africa (5.1%), South America (5.0%), and Oceania (2.4%). Research spans 42 countries (Figure 1), with the USA leading honey bee health studies (26.5%). In Europe, key contributors include Italy (8.9%), the UK (4.7%), Germany (3.2%), and Slovenia (3.1%).

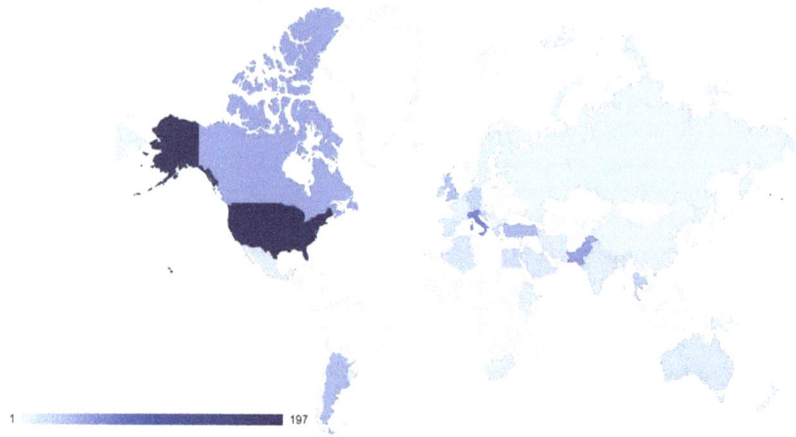

Figure 1. World map showing the density of identified practices. The density of the color increases with the number of identified practice records; light blue represents fewer, dark blue represents more practice records.

To better understand temporal patterns in the implementation of beekeeping practices, we analyzed the seasonality of application across all practice records with available monthly data (Figure 2). We shifted the seasonal data of the Southern Hemisphere at 6 months rate to get a unified calendar (see Section 2). After shifting, globally, the number of practice records peaks between July and October, with a noticeable decline in the winter months forming a seasonal pattern. Since the numbers for the Northern Hemisphere always account for 90% up to 98% of the global numbers, the observed pattern is largely similar. This is probably caused by the high proportion of studies on *Varroa* in the literature. In contrast, the Southern Hemisphere shows a smaller but distinct peak during the months of September to November.

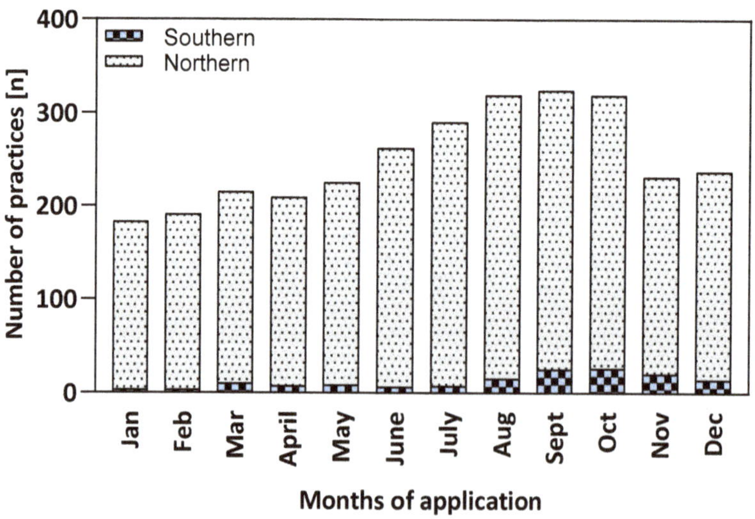

Figure 2. Monthly distribution of identified beekeeping practice records by region. Bars are stacked to show contributions from Northern and Southern hemisphere to the global total (n = 744). Most practice records were documented in the Northern regions.

3.2. Practice Themes and Frequencies

We identified 191 studies resulting in a total of 744 practice records, capturing distinct combinations of practices and outcomes. Practice records were categorized into 8 themes, which are summarized in the Supplemental Material S1. Biosecurity measures accounted for the majority of these practice records (n = 616, 82.8%), while records under good beekeeping practices represented a smaller portion (n = 128, 17.2%). The theme of varroosis was by far the most extensively covered, with 424 practice records (57.0%) addressing this critical issue. General apiary management followed with 128 practice records (17.2%), reflecting its importance in overall colony health and productivity. American foulbrood (AFB) was the third most represented theme with 72 practice records (9.7%), highlighting its continued relevance as a significant bacterial disease in apiculture. Other notable themes included *Tropilaelaps* mite infestation (65 practices, 8.7%) and small hive beetle (aethinosis, 34 practice records, 4.6%). Lesser-studied themes such as european foulbrood (EFB), greater wax moth, and chalkbrood collectively accounted for fewer than 3% of the total practice records, with only 19, 1, and 1 practice records, respectively.

For example, colony management was the most frequent sub-theme (46.9%) for general apiary management. Recent research on *Varroa* mites (varroosis) focuses on "soft" acaricides accounting for 58.5% of practice records, followed by "hard" acaricides (21.0%). With 41.7% antibiotics remain the most important method to combat AFB, followed by 22.2% of biotechnical methods. Studies on treating *Tropilaelaps* spp. focus mainly on "soft" acaricides (81.5%), with biotechnical methods (7.7%) and "hard" acaricides (9.2%) being less common. For the small hive beetle (*Aethina tumida*), in-hive traps dominate (55.9%), followed by hive entrance modifications (17.6%). Table 1 provides a detailed description of the sub-themes and their distribution for general apiary management, varroosis, AFB, tropilaelapsis and aethinosis.

Table 1. Sub-themes ranked by the percentage representation of identified practice records (n = total number of practice records).

General Apiary Management		Varroosis		AFB *		Tropilaelapsis *		Aethinosis *	
n = 128		n = 424		n = 72		n = 65		n = 34	
Colony management	46.9%	"Soft" acaricides	58.5%	Antibiotics	41.7%	"Soft" acaricides	81.5%	In-hive traps	55.9%
Feeding and Watering	35.2%	"Hard" acaricides	21.0%	Biotechnical methods	22.2%	"Hard" acaricides	9.2%	Hive entrance modification	17.6%
Queen management	10.9%	Biotechnical methods	9.7%	Organic agents	12.5%	Biotechnical methods	7.7%	"Soft" acaricides	8.8%
Hygiene	7.0%	Biotechnical methods + "soft" acaricides	6.6%	Laboratory testing	6.9%	Breeding for resistance	1.5%	Colony management	8.8%
		Colony management	3.3%	Colony management	4.2%			Traps outside of hives	5.9%
		Breeding for resistance	0.7%	Breeding for resistance	4.2%			Breeding for resistance	2.9%
		"Soft" acaricides + "hard" acaricides	0.2%	Stamping-out	2.8%				
				Vaccination	2.8%				
				Biotechnical methods + antibiotics	2.8%				

* Total % do not add up to 100% due to rounding to one decimal place.

3.3. General Apiary Management

The most abundant sub-theme was colony management (n = 60). Practice records associated with reduced annual or winter losses include starting new colonies (splits: 11.7%; nucleus colonies: 11.7%), specific hive types (small hives: 10.0%; winter isolation: 3.3%; summer isolation: 1.7%), and swarm control (8.3%). The operational approach, whether organic, conventional, or chemical-free, also plays a significant role, with each type equally represented (6.7%) in the data. Additionally, comb management (13.0%) and the use of domestic queens (8.7%) show positive effects on colony survival. All nine practice records linked to increased honey yields are evenly represented (11.1% each) and are primarily related to colony management strategies (Supplemental Material S1).

According to some researchers, infestation levels (*Varroa*, viruses, *Nosema*) have been positively influenced by the method of starting new colonies (splits 12.0%, nucleus colonies 12.0%), comb hygiene (12.0%), swarming (8.0%), operational type (conventional 8.0%, organic 8.0%), natural forage (8.0%), and less so by annual queen replacement (4.0%) and pollen supplementation (4.0%; Table 2).

Research on queen management primarily focuses on annual queen replacement (42.9%) and its effects on colony losses, productivity, and health. Additional studies compare the performance of domestic (21.4%) versus imported (28.6%) queens (Supplemental Material S1).

Table 2. General apiary management: the significant positive impacts of practice records on honey bees. n = total number of practice records.

***Varroa* + Viruses + *Nosema* Infestation**		**Honey Yields ***		**Annual/Winter Losses + Number of Colonies ***	
n = 25		n = 9		n = 23	
Freezing old brood comb before reuse	12.0%	Annual queen replacement	11.1%	Freezing old brood comb before reuse	13.0%
Starting new colonies by buying nuclei	12.0%	Conventional operation	11.1%	Starting new colonies by buying nuclei	13.0%
Starting new colonies from splits	12.0%	Freezing old brood comb before reuse	11.1%	Starting new colonies from splits	13.0%
Conventional operation	8.0%	Isolated hives in summer	11.1%	Domestic queens	8.7%
Low colony densities	8.0%	Low colony densities	11.1%	Reusing equipment from dead colonies immediately	8.7%
Natural forage	8.0%	Organic operation	11.1%	Annual queen replacement	4.4%
Organic operation	8.0%	Reusing equipment from dead colonies immediately	11.1%	Conventional operation	4.4%
Small hives	8.0%	Starting new colonies by buying nuclei	11.1%	Low colony densities	4.4%
Swarming	8.0%	Starting new colonies from splits	11.1%	Natural forage	4.4%
Annual queen replacement	4.0%			Organic operation	4.4%
Pollen supplementation	4.0%			Pollen supplementation	4.4%
Supplement BEEWELL AminoPlus	4.0%			Small hives	4.4%
Swarming + low colony densities	4.0%			Swarming	4.4%
				Swarming + low colony densities	4.4%
				Wintering strong colonies	4.4%

* Total % do not add up to 100% due to rounding to one decimal place.

Nutrition of the colonies can be influenced by location of an apiary within different landscape contexts, for example, whether it is situated near natural habitats, agricultural fields, or urban areas. These investigations represent 24.4% and 17.8% for natural and agricultural forage, respectively. Supplemental feeding can mean both: providing bees with additional protein or energy sources. Protein sources may be natural pollen (represented with 17.8%), protein substitute (with 13.3%) or amino acid product (4.4%). As a source of energy, traditionally sugar syrup (11.1%) or high fructose corn syrup (11.1%) can be fed to honey bees.

Four practices were linked to higher annual or winter colony losses: chemical-free operations (20.0%), imported queens (40.0%), long-distance transportation (20.0%), and protein supplementation (20.0%). Additionally, lower honey yields were reported for chemical-free operations (50.0%) and shook swarms (50.0%, Supplemental Material S1).

3.4. Varroosis

Varroosis is of main interest in our study, with 424 identified practice records derived from 114 papers. The most extensively researched are "soft" acaricides (n = 248), particularly oxalic acid (36.3%), formic acid (27.8%), and thymol (22.2%). Studies on "hard" acaricides (n = 89) primarily examine tau-fluvalinate (30.3%), amitraz (28.1%), coumaphos (23.6%), and flumethrin (18.0%). Biotechnical methods for *Varroa* control include drone brood removal (22.0%) and hyperthermia (22.0%), queen caging (17.1%), powdered sugar dusting, requeening, and small cell foundation (7.3% each), alongside swarming and total brood removal (4.9% each).

Among the 177 practice records with a positive effect on *Varroa* reduction, oxalic acid was most frequently mentioned (21.5%), followed by other "soft" acaricides such as formic acid (10.7%) and thymol (9.6%). Common "hard" acaricides, including tau-fluvalinate, coumaphos, and amitraz, also proved effective. Biotechnical methods like drone brood removal, queen caging, hyperthermia, and trapping combs were additionally reported. Practices like "queen caging + oxalic acid" and regular mite monitoring were positively associated with both increased honey yields and reduced colony losses (Table 3).

Adverse effects (n = 24) on honey bee colonies have been reported associated with the use of formic acid (37.5%), oxalic acid (29.2%), hyperthermia (12.5%), thymol (12.5%), queen caging (4.2%) and VarroMed (4.2%). Some substances have been reported to leave residues in wax and/or honey (n = 6), namely coumaphos, tau-fluvalinate and thymol. Reduced honey yields (n = 3) have anecdotally been observed with queen caging, hyperthermia and swarming. This highlights the importance of balancing efficacy with potential drawbacks in *Varroa* management strategies.

The seasonal distribution and effectiveness of various practices used to treat varroosis show that the majority of treatments include the use of "soft" acaricides, which were mainly used between July and October, aligning with the period of highest *Varroa* pressure after honey harvest. Biotechnical methods and management strategies are used more evenly throughout the year but increase slightly in late summer (Figure 3a). Significant high treatment success ("positive") of oxalic acid is within a range of 84.6% to 85.7% efficacy most common during broodless periods (winter in the Northern hemisphere). In contrast, treatments applied in early spring (March–May) tend to result in a more varied range of "positive" outcomes between 42.8% to 84.2%, including a higher share of moderate, low positive and even "negative" (7.3–14.3%) effects. This trend is similar for the summer months (June to August) with a maximum of 74.1% "positive" and 8.3% "negative" outcome in June and July, respectively (Figure 3b).

Table 3. Varroosis: the positive impact of practice records on the *Varroa* mite infestation, honey yields and colony losses. Practice records with frequencies below 1% were summarized as "other". n = number of practice records.

Varroa Infestation Levels		Honey Yields *		Colony Losses *	
n = 177		n = 7		n = 15	
Oxalic acid	21.5%	Queen caging + oxalic acid	14.3%	Amitraz	20.0%
Tau-fluvalinate	11.3%	Drone brood removal	14.3%	Monitor monthly and apply miticides when above 3.0 *Varroa* mites/100 bees	20.0%
Formic acid	10.7%	Monitor monthly and apply miticides when above 3.0 *Varroa* mites/100 bees	14.3%	Oxalic acid	20.0%
Thymol	9.6%	Tau-fluvalinate	14.3%	No bee colony import	13.3%
Coumaphos	7.3%	Coumaphos	14.3%	Adult bees mite infestation < 6%	6.7%
Amitraz	5.1%	Formic acid	14.3%	Apply miticides when above *Varroa* 4.0 mites/100 bees	6.7%
Flumethrine	5.1%	Apply miticides when above 2.0 *Varroa* mites/100 bees	14.3%	Brood removal + oxalic acid	6.7%
Queen caging + oxalic acid	5.1%			Queen caging + oxalic acid	6.7%
Drone brood removal	2.8%				
Hop acids	2.3%				
VarroMed	2.3%				
Formic acid + oxalic acid **	1.7%				
Hyperthermia	1.7%				
Trapping comb + oxalic acid	1.1%				
Other	12.4%				

* Total % do not add up to 100% due to rounding to one decimal place. ** applied separately.

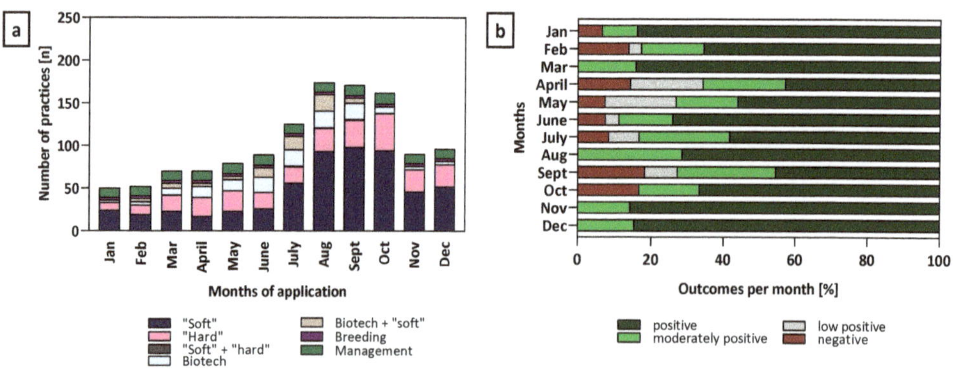

Figure 3. Practices used to treat varroosis. (**a**) categories of *Varroa* mite treatments over the months of application: "soft", "hard" or a combination of "soft" and "hard" acaricides, biotechnical methods or in combination with "soft acaricides, breeding for resistance and colony management practice records. (**b**) relative monthly outcomes of practice records including oxalic acid treatments for *Varroa* control. Bars represent the percentage share of each efficacy outcome category per month (negative, low positive, moderately positive, positive efficacy). n = 424 practice records.

3.5. American Foulbrood

The most research on AFB has been evaluating the use of antibiotics (n = 30). Of these, tylosin has been evaluated most often (53.3%), followed by oxytetracycline (33.3%). Erythromycin and lincomycin was evaluated in a small number of studies.

Practices for treating AFB include shook swarms, primarily for spore reduction (69.2%) and colony recovery (29.1%). Antibiotics such as tylosin and oxytetracycline positively impact the clinical course of the disease but may leave residues in honey. Additional practice records with positive or negative effects related to AFB are summarized in Table 4.

Table 4. AFB: the positive and negative impact of practice records on honey bee colonies. n = number of practice records.

Positive				Negative	
Complete or clinical recovery * n = 55		Number of spores n = 13		Residues in honey * n = 3	
Shook swarm	29.1%	Shook swarm	69.2%	Oxytetracycline	33.3%
Tylosin	21.8%	Requeening with hygienic queens	15.4%	Tylosin	33.3%
Oxytetracycline	14.5%	Low colony densities	7.7%	Erythromycin	33.3%
Cultivation tests from adult bees	5.5%	Propolis	7.7%		
Low colony densities	5.5%				
Burning of clinically infected colonies	3.6%				
Lincomycin	3.6%				
Oral vaccination of queen	3.6%				
Propolis	3.6%				
Requeening with hygienic queens	3.6%				
Cultivation tests from honey	1.8%				
Hygienic behavior	1.8%				
Partial shook swarm + oxytetracycline	1.8%				

* Total % do not add up to 100% due to rounding to one decimal place.

The seasonal distribution of practice records targeting american foulbrood (AFB) reveals a clear peak during spring and early summer, with the highest number of 29 and 36 interventions recorded in April and June, respectively (Figure 4). Among these, the shook swarm method stands out as a targeted strategy used exclusively during the period between February and early summer in June with a minimum of 2 and a maximum of 7 counted practice records, respectively.

3.6. Small Hive Beetle

Two sub-themes dominate as treatment against SHB infestation: in-hive traps (n = 19) and hive entrance modification (n = 6). In-hive traps can incorporate chemical substances such as fipronil (47.4%), coumaphos (15.8%), acetamiprid (5.3%) or slaked lime (5.3%) or a bait (5.3%) or can be constructed as a pitfall (10.5%). Hive entrance can be modified by its upper location (33.3%), by narrowing the entrance to the size of a small pipe (33.3%) or by combination of these two methods (33.3%).

In general, practice records on aethinosis were rather underrepresented 4.6% of 744 records. Positive effect on SHB infestation reduction has been described with the use of in-hive traps with fipronil (36.4%). Further, traps with acetamiprid or coumaphos, baited traps, hive management, hygienic behavior, traps in front of the hive and hive modification have occasionally been evaluated with positive effect on SHB reduction, formic acid, oxalic acid and upper location of the hive entrance seem to have no or little effect on the SHB infestation of the honey bee colonies.

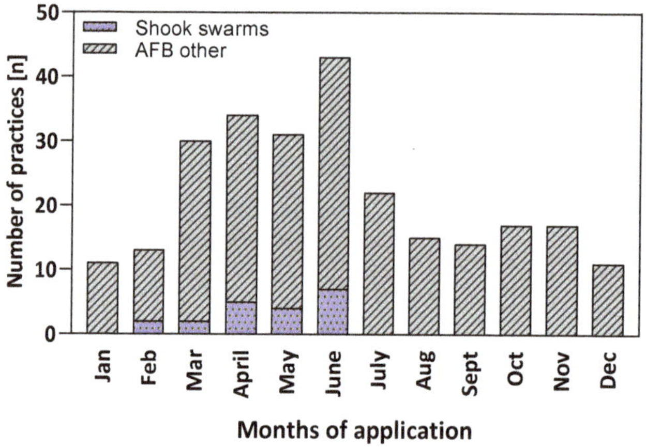

Figure 4. Monthly distribution of practice records targeting american foulbrood (AFB), including shook swarm applications and others. Bars represent the total number of AFB-related practice records per month, with the subset of shook swarm treatments shown in blue (n = 74).

3.7. *Tropilaelaps spp.*

Practice records on *Tropilaelaps* infestations were rather underrepresented with 8.7% of 744 practices, respectively. A detailed list of treatments is found in the Supplemental Table S1.

Similarly to the *Varroa* mite, "soft" acaricides, "hard" acaricides and biotechnical methods are also used to investigate the control of the *Tropilaelaps* mite. Most research is assigned to the sub-theme "soft" acaricides (n = 53) and among these, formic acid (30.2%), thymol (13.2%) and sulfur (11.3%) have been evaluated. Hop acids and essential oils were occasionally studied as well. Classical synthetic varroacides such as flumethrine, amitraz and coumaphos, have rarely been investigated as countermeasures for *Tropilaelaps* spp. (n = 6), as well as some biotechnical methods such as queen caging, queen removal, splitting colonies or powdered sugar dusting. Effective and ineffective treatments for the *Tropilaelaps* mite include formic acid (35.3%), flumethrine (11.8%) and sulfur (11.8%). Insufficient reduction has been briefly observed when investigating hop acids (50.0%), amitraz (25.0%) and sulfur (25.0%).

3.8. *Species and Subspecies Used*

With the exception of one study, all publications investigated honey bee colonies of the species *Apis mellifera* (99.9%). In most cases (73.7%), the subspecies was not further specified. Among those that did provide more detail, *A. m. ligustica* (12.2%) and *A. m. carnica* (6.3%) were the most commonly used. Only a single study (0.1%) was conducted on *Apis cerana cerana* Fabricius, indicating that this species is rarely used in applied field studies worldwide. The remaining records referred either to specific subspecies such as *A. m. intermissa*, *A. m. mellifera*, or hybrid lines like Buckfast, or involved studies where more than one subspecies was used.

4. Discussion

This study presents a structured synthesis of 744 beekeeping practice records from 191 peer-reviewed field studies and offers a comparative framework to assess the efficacy, seasonality, and geographic relevance of these practice records. The intention is to provide a growing foundation for applied apicultural research, targeting beekeepers, extension services, representing people who link research and practice, and policymakers seeking reliable, field-tested strategies to support honey bee health and productivity.

Identifying globally best practices remains difficult due to the strong influence of environmental and climatic conditions, regulatory variations by local governments, regional differences in beekeeping practices, and varying beekeeping philosophies. These factors limit the comparability of practices and outcomes across studies. Despite efforts to achieve comprehensive coverage, manual screening and methodological heterogeneity among studies may have introduced limitations or selection biases. To improve reproducibility and reduce potential selection bias, we recommend applying standardized systematic review methodologies [30,31].

Our study focused on major health-related drivers, namely *Varroa destructor*, american foulbrood (AFB), and general apiary management, while also covering underrepresented but emerging themes such as *Tropilaelaps* spp. and small hive beetle (*Aethina tumida*). We collected scientific data from 42 countries and all continents inhabited by honey bees. The majority of studies included originate from Europe and North America, followed by Asia, Africa, South America and Oceania and the main study object was *A. mellifera* L. Despite broad geographic coverage, practices from the global South remain underrepresented, suggesting the need for more field-based studies that reflect diverse ecological and socioeconomic contexts. In addition, all practice records were categorized by season. *Varroa* treatments were most frequently applied in August and September following honey harvest, while interventions against AFB were concentrated in spring and early summer, during periods of intense brood rearing. Across all themes, most practices were implemented between August and October. The resulting dataset (Supplemental Table S1) provides a foundation for more targeted analyses. Focused, topic-specific reviews, for example, on biotechnical *Varroa* control or seasonal timing of AFB interventions, could build on this work. Here, we present only a few illustrative examples in the form of simple calendars.

Good beekeeping practice (GBP) plays a central role in sustaining profitable and healthy colonies [2]. Furthermore, it also prevents or significantly reduces the impact of honey bee diseases. We identified 128 practice records from 27 peer-reviewed studies in the categories, colony management, queen management, feeding and watering, and hygiene [1].

Several studies had a particularly strong influence on our findings, identifying best management practices (BMPs) that significantly reduced colony losses, pathogen levels and increased profitability [2,32]. Building on this, ref. [33] provided updated survey data on colony losses in the USA, offering critical baseline information to contextualize management practices and their outcomes. Further, ref. [9] developed a tool to identify high-impact practices through data-driven analysis.

Our findings also support the ongoing shift toward organic and sustainable beekeeping, although evidence suggests that abandoning treatments, especially for *Varroa*, altogether may increase colony losses [34]. Instead, the effective use of soft acaricides and biotechnical methods can achieve comparable outcomes, if colony management is adapted to seasonal and local contexts.

The investigations involved in our study uniformly confirm the positive influence of low colony densities within the apiary on the *Varroa* mite infestation level and the colony mortality [35,36]. Further, overwintering, performance and health of the honey bee colonies can be significantly impacted by certain hive modifications, in accordance with the climatic region, for instance in hot summers in arid regions vs. harsh winters in the northern countries [26,27]. Recent findings support such adaptions. Thus, hive insulation with simple covers can reduce food consumption and improve overwintering survival, even in moderate climates [37]. Swarming usually results in economic losses for the beekeeper but can provide natural brood interruption, helping reduce *Varroa* levels. As an alternative, the technique of splitting colonies takes advantage of brood interruption without the negative effects of swarming [38,39].

Queen management is a critical component of colony performance. Frequent queen losses are a known risk factor in the annual losses' statistics [40,41]. Hence proper queen management has been thoroughly studied. Ref. [42] further demonstrated that preparation of queenless colonies for at least five days significantly improves virgin queen acceptance. If the queen replacement was performed with hives being queenless for only one day, the acceptance rate of immature queens was significantly lower. Our findings support existing evidence that younger queens, locally bred queens, and proper requeening timing improve colony outcomes [7,43–46]. Such data-driven recommendations complement our broader results and emphasize the need to adjust requeening strategies to colony condition and timing. Established breeding programs allow to increase the honey production and positively modify the behavior of the honey bee colony.

While mating behavior is recognized as important, it remains understudied in current research. Queens mate with multiple drones, enhancing genetic diversity and resilience, brood viability, and disease resistance [47,48]. This underlines the importance of using well-mated queens and ensuring good mating conditions during rearing. Additionally, recent findings show that genetical resistance traits, such as mite non-reproduction, can be maintained through selective breeding under natural mite pressure [49]. While still emerging, these approaches hold promise for reducing chemical dependency to reduce chemical treatments and improve colony health via genetics-based management [50].

Nutrition operates as a central determinant of colony performance and resilience, mediating interactions between landscape, season, disease susceptibility, and productivity. Studies in our dataset show that pollen diversity, floral richness, and targeted supplemental feeding can significantly enhance immunity, brood development, and overwintering success. Thus, a diverse diet supports colony health [51–53]. However, findings on protein supplements remain rather inconsistent.

As shown in an analysis of 100 years of honey bee research, *Varroa*-related studies became a central focus in the 1990s and remain highly relevant today [30]. Several in-depth reviews have been written that provide a thorough summary of the biology and control of the parasitic mite [14,15,54–56]. We identified 424 practice records from 114 original articles dedicated to *Varroa* control, covering chemical, biotechnical, and integrated pest management (IPM). No universal solution emerged, but multiple effective strategies are available.

Recent trends emphasize Integrated Pest Management (IPM) and reduced reliance on "hard" acaricides, aligning with One Health principles that consider animal, human, and environmental health [1]. Our analysis shows that, under appropriate conditions, biotechnical methods and "soft" acaricides, such as oxalic or formic acid, can be as effective as synthetic treatments. Recent reviews [55,56] emphasize the importance of timing and synergy: combining brood interruption with oxalic acid sublimation often yields best results [57]. Nonetheless, synthetic acaricides continue to play a key role. Depending on the region and infestation level, they remain highly effective and usually do not impair colony development. However, resistance issues and residue accumulation must be considered.

Coumaphos seems to be less effective, probably due to the development of mite resistance. Additionally, coumaphos leaves the most residues in wax and to some degree also in honey [58,59].

Among synthetic acaricides, amitraz remains most commonly used in Europe [60].

Soft acaricides (e.g., oxalic and formic acid) are promising but require precise application depending on brood presence, temperature, and colony size. Sublimation has been occasionally referred to as the most bee-friendly application method of oxalic acid application. Consistent findings across studies confirm that oxalic acid treatment during the broodless period is highly effective, regardless of application method [57,61–65]. Formic acid and thymol are more sensitive to temperature and colony status, complicating general recommendations. Formic acid may cause negative effects like brood loss or queen damage under suboptimal conditions [66–68].

Biotechnical methods vary in efficacy. Brood removal, queen caging, and trapping combs are generally effective [69–71], while sugar dusting and small-cell foundations show limited success [72,73]. Monitoring of mite drop post-treatment is critical to evaluate success, despite treatment method [74].

We identified 72 practice records from 21 studies addressing AFB, categorized into several sub-themes. Although antibiotic use is prohibited in some regions, it remains common elsewhere. Correspondingly, many investigations have been performed on efficacy, dosage and application methods of antibiotics. Traditionally, oxytetracyclines has been widely used, which resulted in resistance. Therefore, new antibiotics, such as tylosin or lincomycin have been tested, with tylosin showing the highest efficacy [75–77]. However, antibiotic residues in honey remain a concern, which can be supported by our results.

To avoid the use of antibiotics, other approaches to combat AFB have been studied. Traditionally, the stamping-out method, including the burning of infected colonies, remains effective but is accompanied by colony and therefore economic losses. The shook swarm method offers a less destructive, yet effective control measure [78,79]. Strategic long-term monitoring and elimination programs have been proven to be successful in fighting AFB [80]. New innovative and still developing approaches may represent the oral vaccination of honey bee queens [81]. The research on European foulbrood is not as extensive. Three methods of combating the pest have been investigated: antibiotics, biotechnical methods, which technically means the shook swarm method and their combination [82,83].

Tropilaelaps spp. mites are a serious threat to *A. mellifera* colonies, particularly in Asia. Compared to *Varroa*, the scientific coverage is limited. The most effective treatment identified in the literature is formic acid evaporation [84,85]. Traditionally, sulfur has been used with reasonable efficacy. Synthetic acaricides may offer additional options, but few studies have evaluated their effectiveness.

Further research is essential to establish effective control strategies for aethinosis. The small list of identified practices reveals that the most promising approach is the use of in-hive traps containing fipronil [86,87]. However, the available studies are scarce and do not cover alternative methods such as external traps, often due to discrepancies between the study design and the criteria for inclusion in this review. As many regions are still SHB-free, preventive strategies were not part of this review but important for further investigation.

5. Conclusions

Taken together, the reviewed practices highlight the necessity of an integrated, multidimensional approach to honey bee colony management. No single intervention is sufficient on its own; rather, effective beekeeping relies on the interplay of multiple, evidence-based strategies. Practices such as regular *Varroa* monitoring and timely treatments, maintaining high-quality and locally adapted queens, adjusting nutritional support to seasonal forage availability, and implementing hygiene and biosecurity protocols form a cohesive management system. These elements reinforce each other; for instance, brood interruption not only aids in *Varroa* control but also aligns with queen replacement strategies, while diversified forage and supplemental feeding can strengthen colonies' resilience against both pathogens and environmental stressors. Our results support the development of a practical framework in which colony health is managed holistically, and it is put in context with season, geography, colony condition, and regulatory environment.

This study provides a structured overview of 744 beekeeping practices from 191 systematically selected field studies, each linked to either a positive or negative impact on colony health or productivity. While the dataset is extensive, we acknowledge that certain gaps remain due to database limitations, inclusion criteria, or underrepresented themes. Still, this framework offers a valuable tool for synthesizing applied honey bee research and for prioritizing practices that have been demonstrated to be successful in real-life beekeeping. Future work should build on this foundation by expanding to additional themes, incorporating long-term and region-specific outcomes, and refining recommendations through ongoing international collaboration. The aim of this work is to promote honey bee health, sustainable beekeeping, and the pollination services essential to ecosystems and agriculture. Therefore, we encourage further validation, contribution, and use of this framework and we support evidence-based decision-making for beekeepers, researchers, veterinarians, and policymakers.

Supplementary Materials: The following supporting information can be downloaded at: https://www.mdpi.com/article/10.3390/insects16101025/s1, Table S1: Supplemental Material S1—How science supports honey bees: Identification of research on best practices in beekeeping.

Author Contributions: Conceptualization, R.B. and V.M.; methodology, R.B.; validation, K.G.; formal analysis, V.M. and K.G.; investigation, V.M.; resources, V.M.; data curation, K.G.; writing—original draft preparation, V.M. and K.G.; writing—review and editing, K.G. and R.B.; visualization, K.G.; supervision, R.B. All authors have read and agreed to the published version of the manuscript.

Funding: This research received no external funding.

Data Availability Statement: Data is contained within the article or Supplementary Material.

Acknowledgments: Open Access Funding by the University of Graz.

Conflicts of Interest: The authors declare no conflicts of interest.

References

1. Rivera-Gomis, J.; Bubnič, J.; Ribarits, A.; Mutinelli, F.; Dahle, B.; Formato, G.; van der Zee, R.; Gracia, M.J.; Soroker, V.; Schuster, S.; et al. Good farming practices in apiculture. *Rev. Sci. Tech.* **2019**, *38*, 1–27. [CrossRef]
2. Tubene, S.; Kulhanek, K.; Rennich, K.; vanEngelsdorp, D. Best management practices increase profitability of small-scale US beekeeping operations. *J. Econ. Entomol.* **2023**, *116*, 47–55. [CrossRef]
3. Pietropaoli, M.; Ribarits, A.; Moosbeckhofer, R.; Köglberger, H.; Alber, O.; Gregorc, A.; Smodis Skerl, M.I.; Presern, J.; Bubnic, J.; Necati Muzet, M.; et al. Biosecurity measures in European beekeeping. *Rev. Sci. Tech.* **2020**, *39*, 725–735. [CrossRef] [PubMed]
4. Antúnez, K.; Invernizzi, C.; Mendoza, Y.; vanEngelsdorp, D.; Zunino, P. Honeybee colony losses in Uruguay during 2013–2014. *Apidologie* **2017**, *48*, 364–370. [CrossRef]
5. Brown, P.; Newstrom-Lloyd, L.E.; Foster, B.J.; Badger, P.H.; McLean, J.A. Winter 2016 honey bee colony losses in New Zealand. *J. Apic. Res.* **2018**, *57*, 278–291. [CrossRef]
6. Gray, A.; Adjlane, N.; Arab, A.; Ballis, A.; Brusbardis, V.; Charrière, J.D.; Chlebo, R.; Coffey, M.F.; Cornelissen, B.; da Costa, C.A.; et al. Honey bee colony winter loss rates for 35 countries participating in the COLOSS survey for winter 2018–2019, and the effects of a new queen on the risk of colony winter loss. *J. Apic. Res.* **2020**, *59*, 744–751. [CrossRef]
7. Oberreiter, H.; Brodschneider, R. Austrian COLOSS survey of honey bee colony winter losses 2018/19 and analysis of hive management practices. *Diversity* **2020**, *12*, 99. [CrossRef]

8. Steinhauer, N.; vanEngelsdorp, D.; Saegerman, C. Prioritizing changes in management practices associated with reduced winter honey bee colony losses for US beekeepers. *Sci. Total Environ.* **2021**, *753*, 141629. [CrossRef]
9. Azzami, K.; Ritter, W.; Tautz, J.; Beier, H. Infection of honey bees with acute bee paralysis virus does not trigger humoral or cellular immune responses. *Arch. Virol.* **2012**, *157*, 689–702. [CrossRef]
10. Evans, J.D.; Spivak, M. Socialized medicine: Individual and communal disease barriers in honey bees. *J. Invertebr. Pathol.* **2010**, *103*, 62–72. [CrossRef]
11. Boecking, O.; Genersch, E. Varroosis—The ongoing crisis in beekeeping. *J. Verbrauch. Lebensm.* **2008**, *3*, 221–228. [CrossRef]
12. Genersch, E.; von der Ohe, W.; Kaatz, H.; Schroeder, A.; Otten, C.; Büchler, R.; Berg, S.; Ritter, W.; Mühlen, W.; Gisder, S.; et al. The German bee monitoring project: A long-term study to understand periodically high winter losses of honey bee colonies. *Apidologie* **2010**, *41*, 232–247. [CrossRef]
13. Rosenkranz, P.; Aumeier, P.; Ziegelmann, B. Biology and control of *Varroa destructor*. *J. Invertebr. Pathol.* **2010**, *103* (Suppl. 1), S96–S119. [CrossRef] [PubMed]
14. Haber, A.I.; Steinhauer, N.A.; vanEngelsdorp, D. Use of chemical and nonchemical methods for the control of *Varroa destructor* (Acari: Varroidae) and associated winter colony losses in US beekeeping operations. *J. Econ. Entomol.* **2019**, *112*, 1509–1525. [CrossRef] [PubMed]
15. Matović, K.; Žarković, A.; Debeljak, Z.; Vidanović, D.; Vasković, N.; Tešović, B.; Ćirić, J. American foulbrood—Old and always new challenge. *Vet. Sci.* **2023**, *10*, 180. [CrossRef]
16. Mosca, M.; Bubnič, J.; Giannetti, L.; Fortugno, L.; Pietropaoli, M.; Manara, V.; Bonerba, E.; Formato, G. Adoption of partial shook swarm in the integrated control of American and European foulbrood of honey bee (*Apis mellifera* L.). *Agriculture* **2023**, *13*, 363. [CrossRef]
17. Burgett, M.; Akratanakul, P.; Morse, R.A. *Tropilaelaps clareae*: A parasite of honeybees in Southeast Asia. *Bee World* **1983**, *64*, 25–28. [CrossRef]
18. Anderson, D.L.; Morgan, M.J. Genetic and morphological variation of bee-parasitic *Tropilaelaps* mites (Acari: Laelapidae): New and re-defined species. *Exp. Appl. Acarol.* **2007**, *43*, 1–24. [CrossRef]
19. Buawangpong, N.; de Guzman, L.I.; Khongphinitbunjong, K.K.; Amanda, M.; Frake, A.M.; Burgett, M.; Chantawannakul, P. Prevalence and reproduction of *Tropilaelaps mercedesae* and *Varroa destructor* in concurrently infested *Apis mellifera* colonies. *Apidologie* **2015**, *46*, 779–786. [CrossRef]
20. Ritter, W., Schneider-Ritter, U. Differences in biology and means of controlling *Varroa jacobsoni* and *Tropilaelaps clareae*, two novel parasitic mites of *Apis mellifera*. In *Africanized Honey Bees and Bee Mites*; Needham, G.R., Page, R.E., Delfinado-Baker, M., Bowman, C.E., Eds.; Ellis Horwood: Chichester, UK, 1988; pp. 383–392.
21. Uzunov, A.; Janashia, I.; Chen, C.; Costa, C.; Kovačić, M. A scientific note on 'Rapid brood decapping'—A method for assessment of honey bee (*Apis mellifera*) brood infestation with *Tropilaelaps mercedesae*. *Apidologie* **2025**, *56*, 40. [CrossRef]
22. Neumann, P.; Ellis, J.D. The small hive beetle (*Aethina tumida* Murray, Coleoptera: Nitidulidae): Distribution, biology, and control of an invasive species. *J. Apic. Res.* **2008**, *47*, 181–183. [CrossRef]
23. Mutinelli, F.; Montarsi, F.; Federico, G.; Granato, A.; Maroni Ponti, A.; Grandinetti, G.; Ferrè, N.; Franco, S.; Duquesne, V.; Rivolta, M.; et al. Detection of *Aethina tumida* Murray (Coleoptera: Nitidulidae) in Italy: Outbreaks and early reaction measures. *J. Apic. Res.* **2014**, *53*, 569–575. [CrossRef]
24. Ellis, J. The honey bee crisis. *Outlooks Pest Manag.* **2012**, *23*, 35–40. [CrossRef]
25. Abou-Shaara, H.F.; Al-Ghamdi, A.A.; Mohamed, A.A. Honey bee colonies performance enhanced by newly modified beehives. *J. Apic. Sci.* **2013**, *57*, 45–57. [CrossRef]
26. Alburaki, M.; Corona, M. Polyurethane honey bee hives provide better winter insulation than wooden hives. *J. Apic. Res.* **2021**, *61*, 190–196. [CrossRef]
27. Meixner, M.D. A historical review of managed honey bee populations in Europe and the United States and the factors that may affect them. *J. Invertebr. Pathol.* **2010**, *103* (Suppl. 1), S80–S95. [CrossRef]
28. Goulson, D.; Nicholls, E.; Botías, C.; Rotheray, E.L. Bee declines driven by combined stress from parasites, pesticides, and lack of flowers. *Science* **2015**, *347*, 1255957. [CrossRef]
29. Hassler, E.E.; Cazier, J.A.; Hopkins, B.; Wilkes, J.T.; Smith, K.; Rünzel, M. A century of discovery: Mining 100 years of honey bee research. *J. Apic. Res.* **2020**, *60*, 3–12. [CrossRef]
30. Haddaway, N.R.; Woodcock, P.; Macura, B.; Collins, A. Making literature reviews more reliable through application of lessons from systematic reviews. *Conserv. Biol.* **2015**, *29*, 1596–1605. [CrossRef]
31. Kulhanek, K.; Steinhauer, N.; Wilkes, J.; Wilson, M.; Spivak, M.; Sagili, R.R.; Tarpy, D.R.; McDermott, E.; Garavito, A.; Rennich, K.; et al. Survey-derived best management practices for backyard beekeepers improve colony health and reduce mortality. *PLoS ONE* **2021**, *16*, e0245490. [CrossRef]
32. Aurell, D.; Bruckner, S.; Wilson, M.; Steinhauer, N.; Williams, G.R. A national survey of managed honey bee colony losses in the USA: Results from the Bee Informed Partnership for 2020–21 and 2021–22. *J. Apic. Res.* **2023**, *63*, 1–14. [CrossRef]

33. Underwood, R.M.; Lawrence, B.L.; Turley, N.E.; Cambron-Kopco, L.D.; Kietzman, P.M.; Traver, B.E.; López-Uribe, M.M. A longitudinal experiment demonstrates that honey bee colonies managed organically are as healthy and productive as those managed conventionally. *Sci. Rep.* **2023**, *13*, 6072. [CrossRef] [PubMed]
34. Frey, E.; Rosenkranz, P. Autumn invasion rates of *Varroa destructor* (Mesostigmata: Varroidae) into honey bee (Hymenoptera: Apidae) colonies and the resulting increase in mite populations. *J. Econ. Entomol.* **2014**, *107*, 508–515. [CrossRef] [PubMed]
35. Dynes, T.L.; Berry, J.A.; Delaplane, K.S.; Brosi, B.J.; de Roode, J.C. Reduced density and visually complex apiaries reduce parasite load and promote honey production and overwintering survival in honey bees. *PLoS ONE* **2019**, *14*, e0216286. [CrossRef] [PubMed]
36. St. Clair, A.L.; Beach, N.J.; Dolezal, A.G. Honey bee hive covers reduce food consumption and colony mortality during overwintering. *PLoS ONE* **2022**, *17*, e0266219. [CrossRef]
37. Guler, A. The effects of the shook swarm technique on honey bee (*Apis mellifera* L.) colony productivity and honey quality. *J. Apic. Res.* **2008**, *47*, 27–34. [CrossRef]
38. Loftus, J.C.; Smith, M.L.; Seeley, T.D. How honey bee colonies survive in the wild: Testing the importance of small nests and frequent swarming. *PLoS ONE* **2016**, *11*, e0150362. [CrossRef]
39. vanEngelsdorp, D.; Hayes, J., Jr.; Underwood, R.M.; Pettis, J.S. A survey of honey bee colony losses in the United States, fall 2008 to spring 2009. *J. Apic. Res.* **2010**, *49*, 7–14. [CrossRef]
40. Kulhanek, K.; Steinhauer, N.; Rennich, K.; Caron, D.M.; Sagili, R.R.; Pettis, J.S.; Ellis, J.D.; Wilson, M.E.; Wilkes, J.T.; Tarpy, D.R.; et al. A national survey of managed honey bee 2015–2016 annual colony losses in the USA. *J. Apic. Res.* **2017**, *56*, 328–340. [CrossRef]
41. Smilga-Spalvina, A.; Spalvins, K.; Veidenbergs, I. Field study: Factors influencing virgin queen bee acceptance rate in *Apis mellifera* colonies. *Czech J. Anim. Sci.* **2024**, *69*, 155–164. [CrossRef]
42. Akyol, E.; Yeninar, H.; Karatepe, M.; Karatepe, B.; Özkök, D. Effects of queen ages on *Varroa destructor* infestation level in honey bee (*Apis mellifera caucasica*) colonies and colony performance. *Ital. J. Anim. Sci.* **2007**, *6*, 143–149. [CrossRef]
43. Giacobino, A.; Molineri, A.; Cagnolo, N.B.; Merke, J.; Orellano, E.; Bertozzi, E.; Masciangelo, G.; Pietronave, H.; Pacini, A.; Salto, C. Queen replacement: The key to prevent winter colony losses in Argentina. *J. Apic. Res.* **2016**, *55*, 335–341. [CrossRef]
44. Kovačić, M.; Puškadija, Z.; Dražić, M.M.; Uzunov, A.; Meixner, M.D.; Büchler, R. Effects of selection and local adaptation on resilience and economic suitability in *Apis mellifera* carnica. *Apidologie* **2020**, *51*, 1062–1073. [CrossRef]
45. Holmes, L.A.; Ovinge, L.P.; Kearns, J.D.; Ibrahim, A.; Wolf Veiga, P.; Guarna, M.M.; Pernal, S.F.; Hoover, S.E. Queen quality, performance, and winter survival of imported and domestic honey bee queen stocks. *Sci. Rep.* **2023**, *13*, 17273. [CrossRef] [PubMed]
46. Rousseau, A.; Fournier, V.; Giovenazzo, P. *Apis mellifera* (Hymenoptera: Apidae) drone sperm quality in relation to age, genetic line, and time of breeding. *Can. Entomol.* **2015**, *147*, 702–711. [CrossRef]
47. Delaney, D.A.; Keller, J.J.; Caren, J.R.; Tarpy, D.R. The physical, insemination, and reproductive quality of honey bee queens (*Apis mellifera* L.). *Apidologie* **2011**, *42*, 1–13. [CrossRef]
48. Guichard, M.; Dietemann, V.; Neuditschko, M.; Dainat, B. Advances and perspectives in selecting resistance traits against the parasitic mite *Varroa destructor* in honey bees. *Genet. Sel. Evol.* **2020**, *52*, 71. [CrossRef]
49. Cobey, S.; Tarpy, D.R.; Woyke, J. Standard methods for instrumental insemination of *Apis mellifera* queens. *J. Apic. Res.* **2013**, *52*, 1–18. [CrossRef]
50. Brodschneider, R.; Crailsheim, K. Nutrition and health in honey bees. *Apidologie* **2010**, *41*, 278–294. [CrossRef]
51. DeGrandi-Hoffman, G.; Chen, Y.; Rivera, R.; Carroll, M.; Chambers, M.; Hidalgo, G.; de Jong, E.W. Honey bee colonies provided with natural forage have lower pathogen loads and higher overwinter survival than those fed protein supplements. *Apidologie* **2016**, *47*, 186–196. [CrossRef]
52. Branchiccela, B.; Castelli, L.; Corona, M.; Díaz-Cetti, S.; Invernizzi, C.; Martínez de la Escalera, G.; Mendoza, Y.; Santos, E.; Silva, C.; Zunino, P.; et al. Impact of nutritional stress on honeybee colony health. *Sci. Rep.* **2019**, *9*, 10156. [CrossRef]
53. Gregorc, A.; Sampson, B. Diagnosis of *Varroa* mite (*Varroa destructor*) and sustainable control in honey bee (*Apis mellifera*) colonies—A review. *Diversity* **2019**, *11*, 243. [CrossRef]
54. Bubnič, J.; Moosbeckhofer, R.; Prešern, J.; Moškrič, A.; Formato, G.; Pietropaoli, M.; Gregorc, A.; Muz, M.N.; Maja Ivana Smodiš Škerl, M.I. Three pillars of *Varroa* control. *Apidologie* **2021**, *52*, 1–29. [CrossRef]
55. Jack, C.J.; Ellis, J.D. Integrated pest management control of *Varroa destructor* (Acari: Varroidae), the most damaging pest of *Apis mellifera* L. (Hymenoptera: Apidae) colonies. *J. Insect Sci.* **2021**, *21*, 6. [CrossRef]
56. Berry, J.A.; Bartlett, L.J.; Bruckner, S.; Baker, C.; Kris Braman, S.; Delaplane, K.S.; Williams, G.R. Assessing repeated oxalic acid vaporization in honey bee (*Apis mellifera*) colonies for control of the ectoparasitic mite *Varroa destructor*. *J. Insect Sci.* **2022**, *22*, 15. [CrossRef] [PubMed]
57. Wallner, K. Varroacides and their residues in bee products. *Apidologie* **1999**, *30*, 235–248. [CrossRef]

58. Martel, A.C.; Zeggane, S.; Aurières, C.; Drajnudel, P.; Faucon, J.P.; Aubert, M. Acaricide residues in honey and wax after treatment of honey bee colonies with Apivar® or Asuntol®50. *Apidologie* **2007**, *38*, 534–544. [CrossRef]
59. Brodschneider, R.; Schlagbauer, J.; Arakelyan, I.; Ballis, A.; Brus, J.; Brusbardis, V.; Cadahía, L.; Charrière, J.-D.; Chlebo, R.; Coffey, M.F. Spatial clusters of *Varroa destructor* control strategies in Europe. *J. Pest Sci.* **2023**, *96*, 759–783. [CrossRef]
60. Gregorc, A.; Planinc, I. Acaricidal effect of oxalic acid in honeybee (*Apis mellifera*) colonies. *Apidologie* **2001**, *32*, 333–340. [CrossRef]
61. Gregorc, A.; Planinc, I. The control of *Varroa destructor* using oxalic acid. *Vet. J.* **2002**, *163*, 306–310. [CrossRef]
62. Rademacher, E.; Harz, M. Oxalic acid for the control of varroosis in honey bee colonies—A review. *Apidologie* **2006**, *37*, 98–120. [CrossRef]
63. Al Toufailia, H.; Scandian, L.; Ratnieks, F.L. Towards integrated control of *Varroa*: 2) Comparing application methods and doses of oxalic acid on the mortality of phoretic *Varroa destructor* mites and their honey bee hosts. *J. Apic. Res.* **2015**, *54*, 108–120. [CrossRef]
64. Jack, C.J.; van Santen, E.; Ellis, J.D. Determining the dose of oxalic acid applied via vaporization needed for the control of the honey bee (*Apis mellifera*) pest *Varroa destructor*. *J. Apic. Res.* **2021**, *60*, 414–420. [CrossRef]
65. Satta, A.; Floris, I.; Eguaras, M.; Cabras, P.; Garau, V.L.; Melis, M. Formic acid-based treatments for control of *Varroa destructor* in a Mediterranean area. *J. Econ. Entomol.* **2005**, *98*, 267–273. [CrossRef] [PubMed]
66. Pietropaoli, M.; Formato, G. Acaricide efficacy and honey bee toxicity of three new formic acid-based products to control *Varroa destructor*. *J. Apic. Res.* **2019**, *58*, 824–830. [CrossRef]
67. Hendriksma, H.P.; Cornelissen, B.; Panziera, D. Liquid and solid matrix formic acid treatment comparison against Varroa mites in honey bee colonies. *J. Apic. Res.* **2023**, *62*, 320–328. [CrossRef]
68. Giacomelli, A.; Pietropaoli, M.; Carvelli, A.; Iacoponi, F.; Formato, G. Combination of thymol treatment (Apiguard®) and caging the queen technique to fight *Varroa destructor*. *Apidologie* **2016**, *47*, 606–616. [CrossRef]
69. Büchler, R.; Uzunov, A.; Kovačić, M.; Prešern, J.; Pietropaoli, M.; Hatjina, F.; Pavlov, B.; Charistos, L.; Formato, G.M.; Galarza, E.; et al. Summer brood interruption as an integrated management strategy for effective *Varroa* control in Europe. *J. Apic. Res.* **2020**, *59*, 764–773. [CrossRef]
70. Mortensen, A.N.; Jochym, M.; Dearden, P.K.; Sainsbury, J.P. Old strategies as modern solutions: Shook swarming reduced *Varroa destructor* infestation rates of managed honey bee (*Apis mellifera*) colonies. *J. Apic. Res.* **2023**, *62*, 1–7. [CrossRef]
71. Ellis, A.M.; Hayes, G.W.; Ellis, J.D. The efficacy of small cell foundation as a varroa mite (*Varroa destructor*) control. *Exp. Appl. Acarol.* **2009**, *47*, 311–316. [CrossRef]
72. Berry, J.A.; Afik, O.; Nolan, M.P.; Delaplane, K.S. Revisiting powdered sugar for Varroa control on honey bees (*Apis mellifera* L.). *J. Apic. Res.* **2012**, *51*, 367–368. [CrossRef]
73. Morfin, N.; Foster, L.J.; Guzman-Novoa, E.; Westendorp, P.V.; Currie, R.W.; Higo, H. *Varroa destructor* economic injury levels and pathogens associated with colony losses in Western Canada. *Front. Bee Sci.* **2024**, *2*, 1355401. [CrossRef]
74. Peng, C.Y.S.; Mussen, E.; Fong, A.; Cheng, P.; Wong, G.; Montague, M.A. Laboratory and field studies on the effects of the antibiotic tylosin on honey *bee Apis mellifera* L. (Hymenoptera: Apidae) development and prevention of American foulbrood disease. *J. Invertebr. Pathol.* **1996**, *67*, 65–71. [CrossRef] [PubMed]
75. Alippi, A.M.; Albo, G.N.; Leniz, D.; Rivera, I.; Zanelli, M.L.; Roca, A.E. Comparative study of tylosin, erythromycin, and oxytetracycline to control American foulbrood of honey bees. *J. Apic. Res.* **1999**, *38*, 149–158. [CrossRef]
76. Pettis, J.S.; Feldlaufer, M.F. Efficacy of lincomycin and tylosin in controlling American foulbrood in honey bee colonies. *J. Apic. Res.* **2005**, *44*, 106–108. [CrossRef]
77. Hansen, H.; Brødsgaard, C.J. Control of American foulbrood by the shaking method. *Apiacta* **2003**, *38*, 140–145.
78. Pernal, S.F.; Albright, R.L.; Melathopoulos, A.P. Evaluation of the shaking technique for the economic management of American foulbrood disease of honey bees (Hymenoptera: Apidae). *J. Econ. Entomol.* **2008**, *101*, 1095–1104. [CrossRef]
79. Locke, B.; Low, M.; Forsgren, E. An integrated management strategy to prevent outbreaks and eliminate infection pressure of American foulbrood disease in a commercial beekeeping operation. *Prev. Vet. Med.* **2019**, *167*, 48–52. [CrossRef]
80. Dickel, F.; Bos, N.M.P.; Hughes, H.; Martín-Hernández, R.; Higes, M.; Kleiser, A.; Freitak, D. The oral vaccination with *Paenibacillus larvae* bacterin can decrease susceptibility to American foulbrood infection in honey bees—A safety and efficacy study. *Front. Vet. Sci.* **2022**, *9*, 946237. [CrossRef]
81. Thompson, H.M.; Waite, R.J.; Wilkins, S.; Brown, M.A.; Bigwood, T.; Shaw, M.; Ridgway, C.; Sharman, M. Effects of shook swarm and supplementary feeding on oxytetracycline levels in honey extracted from treated colonies. *Apidologie* **2006**, *37*, 51–57. [CrossRef]
82. Mosca, M.; Gyorffy, A.; Pietropaoli, M.; Giannetti, L.; Cersini, A.; Fortugno, L.; Formato, G. IPM strategy to control EFB in *Apis mellifera*: Oxytetracycline treatment combined with partial shook swarm and queen caging. *Vet. Sci.* **2024**, *11*, 28. [CrossRef]
83. Mahmood, R.; Wagchoure, E.S.; Raja, S.; Sarwar, G.; Aslam, M. Effect of thymol and formic acid against ectoparasitic brood mite *Tropilaelaps clareae* in *Apis mellifera* colonies. *Pak. J. Zool.* **2011**, *43*, 41–46.
84. Pettis, J.S.; Rose, R.; Chaimanee, V. Chemical and cultural control of *Tropilaelaps mercedesae* mites in honeybee (*Apis mellifera*) colonies in Northern Thailand. *PLoS ONE* **2017**, *12*, e0188063. [CrossRef]

85. Levot, G.W. An insecticidal refuge trap to control adult small hive beetle, *Aethina tumida* Murray (Coleoptera: Nitidulidae) in honey bee colonies. *J. Apic. Res.* **2008**, *47*, 222–228. [CrossRef]
86. Levot, G.W.; Somerville, D. Efficacy and safety of the insecticidal small hive beetle refuge trap APITHOR™ in bee hives. *Aust. J. Entomol.* **2012**, *51*, 198–204. [CrossRef]
87. Kleckner, K.; De Carolis, A.; Jack, C.; Stuhl, C.; Formato, G.; Ellis, J.D. A novel acute toxicity bioassay and field trial to evaluate compounds for small hive beetle control. *Appl. Sci.* **2022**, *12*, 9905. [CrossRef]

insects

MDPI

Article

Toxicity of Consecutive Treatments Combining Synthetic and Organic Miticides to Nurse Bees of *Apis mellifera*

HeeJin Kim [1], Euijin You [2], JooHeon Cha [3], Si Hyeock Lee [4] and Young Ho Kim [1,2,3,*]

[1] Department of Ecological Science, Kyungpook National University, Sangju 37224, Gyeongbuk, Republic of Korea; maybe8541@naver.com
[2] Research Institute of Invertebrate Vector, Kyungpook National University, Sangju 37224, Gyeongbuk, Republic of Korea; yej8415@naver.com
[3] Department of Vector Entomology, Kyungpook National University, Sangju 37224, Gyeongbuk, Republic of Korea; 19cjh95@gmail.com
[4] Department of Agricultural Biotechnology, Seoul National University, Seoul 08826, Republic of Korea; shlee22@snu.ac.kr
* Correspondence: yhkim05@knu.ac.kr

Simple Summary: Ectoparasitic mites pose serious threats to honey bee health and contribute to global colony losses. To manage mites, beekeepers commonly use both synthetic and organic miticides, sometimes in succession. While many studies focus on forager bees, nurse bees—who stay inside the hive—are also at risk due to direct contact and prolonged exposure to residual chemicals. This study investigated the toxic effects of single and consecutive applications of three synthetic (fluvalinate, coumaphos, and amitraz) and two organic (formic acid and oxalic acid) miticides, at field-realistic concentrations, on nurse bees. The results showed that synthetic miticides caused minimal acute mortality under the conditions tested, while organic miticides—especially formic acid—significantly reduced survival. In most cases, consecutive treatments did not worsen toxicity beyond that of the corresponding organic miticide alone, suggesting that the organic agents were the primary drivers of mortality. Overall, this study highlights the potential risks of combining miticides and underscores the need for safer application strategies to protect vulnerable in-hive bees like nurse bees.

Abstract: *Varroa destructor*, an ectoparasitic mite of honey bees, is a major contributor to global colony declines. To manage infestations, beekeepers frequently apply both synthetic and organic miticides—sometimes in combination. While much attention has been given to foragers' pesticide exposure, nurse bees are also at risk due to direct miticide contact and prolonged exposure to residuals in hive matrices. Despite this, little is known about the effects of sequential miticide applications on nurse bees. In this study, we evaluated the toxicities of single and pairwise consecutive applications of three synthetic miticides (fluvalinate, coumaphos, and amitraz) and two organic miticides (formic acid and oxalic acid) at field-realistic concentrations. Miticides were topically applied to nurse bees with 24 h between consecutive treatments. Synthetic miticides caused minimal mortality, while both organic miticides—especially formic acid—significantly reduced survivorship. Consecutive treatments generally produced mortality rates comparable to that of the corresponding organic miticide alone, indicating that observed toxicities were primarily driven by the organic agents. Notably, the fluvalinate–formic acid combination induced no significant adverse effects. These findings highlight the need for caution when using miticides in combination and support the development of safer application strategies to protect vulnerable in-hive bee populations..

 https://doi.org/10.3390/insects16070657

Keywords: *Apis mellifera*; nurse bee; synthetic miticide; organic miticide; toxicity; *Varroa destructor*

1. Introduction

The western honey bee, *Apis mellifera*, is a key pollinator essential for both agricultural productivity and the maintenance of natural ecosystems. In addition, honey bees contribute significantly to beekeeping industries by producing valuable products, such as honey, propolis, and royal jelly. Despite their importance, global honey bee populations have been declining in recent years. Previous studies have reported that over 20% of honey bee colonies disappeared in the United States between 2018 and 2019 and in European countries between 2019 and 2020 [1,2]. This is a similar problem in the Republic of Korea, where severe colony losses were repeatedly reported during overwintering in 2021 and 2022 [3]. Various factors have been implicated in honey bee colony collapse, including infestations by ectoparasites, the spread of diseases, and abnormal climate conditions [3–5]. Among these, *Varroa destructor*, an ectoparasitic mite, is widely recognized as a major threat to honey bee health. Studies have demonstrated that *V. destructor* infection weakens the immune system, increases the susceptibility to viral infections, and inhibits the development of the honey bees, leading to colony collapse [6–9].

To control *V. destructor*, beekeepers worldwide employ various synthetic miticides, such as tau-fluvalinate (Flu), coumaphos (Cou), and amitraz (Ami), as well as organic miticides, including formic acid (FA) and oxalic acid (OA) [10,11]. Flu, Cou, and Ami exert neurotoxic effects on *V. destructor* by targeting voltage-gated sodium channels, acetylcholinesterase, and octopamine β receptors, respectively, while having relatively low toxicity to honey bees [12,13]. Among the organic miticides, FA acts by inhibiting cytochrome c oxidase [14], whereas the mode of action of OA remains unclear.

Despite their efficacy, the improper application of miticides can have detrimental effects on individual honey bees and entire colonies [3,15]. For example, Flu exposure has been shown to increase honey bees' susceptibility to deformed wing virus [16], disrupt foraging behaviors [17,18], and impair queen reproductive capacity [19]. Similarly, Cou has been reported to reduce queen fertility and interfere with trophallaxis among workers [20,21]. Furthermore, the misuse of Ami has been associated with larval cell death and the inhibition of cuticle protease activity in worker bees [22,23].

Organic miticides can also cause significant harm if misapplied. Exposure to FA has been linked to cell death, respiratory interference, queen loss, and brood reductions [23–27]. Similarly, excessive OA use has been reported to incur cell death, reduced lifespans, and brood removal [7,26,28]. These findings underscore the risks associated with improper miticide usage in apiculture.

Recently, the emergence of *V. destructor* populations resistant to synthetic miticides, including Flu, Cou, and Ami, has been reported worldwide [5,29,30]. In the Republic of Korea, Flu-resistant *V. destructor* populations have rapidly increased since 2021 [31]. The widespread resistance of mites to synthetic miticides has made mite control increasingly difficult and has contributed to further honey bee losses. In response, many beekeepers have, without adequate knowledge of resistance dynamics, resorted to increasing miticide applications. Furthermore, as synthetic miticides alone are no longer sufficient for mite control, the simultaneous use of synthetic and organic miticides has become a common practice.

The continuous application of miticides leads to their accumulation in hive matrices, such as pollen and wax [10,32]. According to previous studies, when Checkmite® strips containing Cou were used in colonies, the concentration of Cou significantly increased in wax [33,34]. In addition, colonies treated with commercial Ami products exhibited high Ami concentrations in their wax and bee bread when compared to untreated colonies [35]. Furthermore, Murcia-Morales et al. [36] reported that when miticides were administered to colonies, each miticide was detected in honey bees and colonies at high concentrations and remained present for weeks. In addition to bee products, various miticides were detected from living or dead honey bees, including adult bees and brood [32,37–39]. Calatayud-Vernich et al. [40] found miticide-contaminated honey bee carcasses in collapsed colonies. Together, these studies indicate that nurse bees are continuously exposed to synthetic miticides that persist inside the hive after application.

Furthermore, given that organic miticides are often introduced into hives where synthetic miticide residues persist and considering that both types of miticides are sometimes applied simultaneously—primarily exposing nurse bees rather than foragers—it is crucial to investigate their combined toxic effects on nurse bees. Therefore, in this study, we assessed the individual toxicity of commonly used synthetic (Flu, Cou, and Ami) and organic (OA and FA) miticides at field-realistic concentrations [41] to nurse bees. Then, we evaluated the synergistic toxicity resulting from sequential treatments involving synthetic and organic miticides, comparing the effects of single and combined treatments to identify miticide combinations that elicit unexpected toxic interactions in nurse bees.

2. Materials and Methods

2.1. Honey Bee Samples

Western honey bee nurse bees, identified based on their cell-cleaning and brood-care behaviors [42], were collected from the experimental apiary of Kyungpook National University in Sangju-si, Gyeongsanbuk-do, Republic of Korea. To enrich nurse bees, we collected individuals exclusively from the central brood frames during peak nursing hours (09:00–14:00), when brood care behaviors are most actively expressed. This practical approach, which has been widely adopted in previous studies [42], allows for non-lethal selection of nurse bees based on temporal task allocation and behavioral cues. To prevent unintended miticide exposure, no synthetic or organic miticides had ever been applied to the experimental colonies. Instead, the powdered sugar method was used for *V. destructor* control [43]. The collected nurse bees were maintained in an incubator (JSBI-250C, JS Research Inc., Gongju-si, Republic of Korea) at 32 °C under dark conditions and were provided with a 50% sucrose solution ad libitum.

2.2. Determination of the Experimental Miticide Concentrations

Analytical standard-grade synthetic miticides, including Flu, Cou, and Ami, were purchased from Sigma Aldrich (Saint Louis, MO, USA). Pure organic miticides, including FA and OA, were obtained from the National Institute of Agricultural Sciences of the Rural Development Administration (RDA, Jeonju-si, Republic of Korea).

To determine experimental concentrations reflective of field applications, we referred to previous studies [44–47] and the manufacturers' manuals of commercial products registered with the Animal and Plant Quarantine Agency (APQA, Gimcheon-si, Republic of Korea) and the Environmental Protection Agency (EPA, Washington, DC, USA) (Table 1). For Flu, we referenced Apistan® (Vita Bee Health Ltd., Basingstoke, UK), a product containing Flu as the active ingredient, since commercial products commonly used in Korea, such as Hongsabang® (Shanxi Hengda Leiao Bio-Technology Co., Ltd.; Yuncheng, China) and Manpu Gold® (Shanxi Weipeng Pharmaceutical Co., Ltd.; Yuncheng, China), have been less extensively studied. According to previous studies [44,46], Apistan® strips release approximately 0.125 μg of Flu per day when applied to a honey bee colony following the manufacturer's instructions. For Cou and Ami, we considered commercial products applied to honey bee colonies via spraying. The undiluted solutions of Coumaking® (Daehan Nupharm, Hwasung-si, Republic of Korea) and Couma-H® (Beesen Bio, Geumsan-gun, Republic of Korea) contain 32 g of Cou, and Soksal-Gold Solution® (Shanxi Weipeng Pharmaceutical Co., Ltd.) and Dr+ Bee (Biobee, Hwaseong-si, Republic of Korea) contain 125 g of Ami. However, when diluted according to the manufacturers' guidelines, the final concentrations of Cou and Ami in the prepared solutions are 0.62 μg/μL and 0.125 μg/μL, respectively. For organic miticides, the experimental concentration of FA was based on a previous study [47], which estimated that honey bees could be exposed to approximately 3 μL of FA per bee over 24 h when 60% FA was applied to a honey bee colony. For OA, since no standardized application method has been established in Korea, we referenced the concentration used in the dribble method, as indicated on product labels registered with the EPA.

Table 1. Information on the miticides tested in this study and their field-realistic concentrations.

Chemical		Product (A.I. %) [a]	Estimated Exposure Concentration	References
Synthetic miticide	Fluvalinate	Apistan (824 mg/strip)	41.67 ppm [b]	[44–46]
	Coumaphos	Coumaking (3.2%)	626.67 ppm [c]	Product label registered at the APQA
		Couma-H (3.2%)	626.67 ppm [c]	
	Amitraz	Soksal-gold Solution (12.5%)	125 ppm [c]	
		Dr+ Bee Solution (12.5%)	125 ppm [c]	
Organic miticide	Formic acid	Handmade solution	600,000 ppm [b]	[47]
	Oxalic acid	Handmade solution	35,000 ppm [c]	Product label registered at the EPA

[a] Active ingredient (%) in the commercial products; [b] estimated field-realistic exposure concentration based on the previous studies; [c] Recommended concentration (ppm) when the application solution is made following the guidelines registered with the Animal and Plant Quarantine Agency (APQA, Republic of Korea) or Environmental Protection Agency (EPA, USA).

2.3. *Bioassay*

To assess the toxicity of sequential miticide treatments, three synthetic miticides (Flu, Cou, and Ami) and two organic miticides (OA and FA) were applied to nurse bees at field-realistic concentrations. Synthetic and then organic miticides were administered sequentially, with a 24 h interval between treatments. Considering that certain miticides, such as Flu, FA, and OA, are applied through evaporation or fumigation, it was essential to ensure exposure to all body tissues. Given that the surface area of a honey bee is approximately 3 cm^2 [48], 1 μL of each prepared miticide solution was topically applied to three body regions on each nurse bee: the head, thorax, and abdomen.

In the first treatment, nurse bees received 1 μL of either acetone (control) or a synthetic miticide in acetone at a specific concentration: 41.67 ppm for Flu, 626.67 ppm for Cou, and 125 ppm for Ami. Following treatment, the honey bees were maintained at 32 °C in darkness for 24 h in an incubator (JS Research Inc.) with ad libitum access to a 50% (*w*/*v*) sucrose solution. After 24 h, the second treatment was administered to the surviving nurse bees. Organic miticides were applied using the same method, with concentrations of 600,000 ppm for FA and 35,000 ppm for OA (Table 1). The treated nurse bees were then maintained under the same conditions used in the first 24 h (Figure 1).

The experimental design included sequential treatment groups in which nurse bees were treated with a synthetic miticide followed by an organic miticide. In the single-treatment groups, only one miticide type was applied, and pure acetone was used in place of the unused miticide. Thus, in the synthetic miticide single-treatment groups, nurse bees were treated with the synthetic miticide followed by acetone, while in the organic miticide single-treatment groups, acetone was applied first, followed by the organic miticide. The control group received acetone in both treatments.

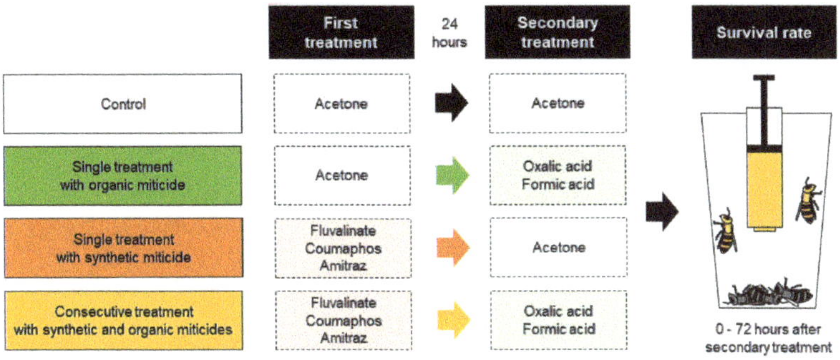

Figure 1. An overview of the experimental treatment procedure. Each group was administered with acetone and/or miticide as described in the left or right dotted boxes during first or secondary treatment, respectively, with a one-day interval. Organic miticides were categorized in light green, dotted box, whereas synthetic miticides were grouped in light orange, dotted box.

Survival rates were recorded 0, 3, 6, 9, 12, 24, 36, 48, 60, and 72 h after the second miticide treatment. Honey bees were considered dead if they exhibited no movement upon stimulation, after which their carcasses were removed from the cages. All experiments were conducted in triplicate, with each experimental replicate performed at least three times.

2.4. Data Analysis

All survival rates from both single and sequential treatments were analyzed using SPSS 27 for Windows (IBM, Armonk, NY, USA). Kaplan–Meier survival analyses using log-rank test pairwise over strata were performed to assess overall survival differences among treatment groups. Survival rates at each time point were calculated as means ± standard errors (SEs) and compared using independent *t*-tests to determine statistical significance. Additionally, lethal time (LT) values, including LT_5, LT_{10}, and LT_{30}, were estimated by extrapolation using Excel (Office 365, Microsoft, Redmond, WA, USA) following the method described by Mesa et al. [49], with slight modifications.

3. Results

3.1. Single Treatments with Synthetic Miticides

When evaluating the toxicity of single treatments with synthetic miticides at field concentrations, we found that the survival trends of nurse bees treated with Flu, Cou, and Ami were statistically similar to those of the control group across the entire 72 h post-treatment (hpt) period ($p > 0.05$) (Table S1 and Figure 2).

Based on the time-dependent data, the survival rate of the Flu group (100.0 ± 0.0%) was identical to that of the control group (100.0 ± 0.0%) at 24 hpt, while nurse bees exposed to Ami and Cou exhibited slightly lower but comparable survival rates of 96.9 ± 3.1% and 95.8 ± 4.2%, respectively (Table 2). At 48 hpt, survival rates remained high, with rates of 100.0 ± 0.0%, 95.8 ± 4.2%, 96.9 ± 3.1%, and 97.5 ± 2.5% recorded for the Flu, Cou, Ami, and control groups, respectively. Although the survival rates of nurse bees receiving the three synthetic miticides treatments were not statistically different at 72 hpt, nurse bees exposed to the miticides exhibited slightly higher survival than those of the control group, with Flu resulting in the highest (96.9 ± 3.1%), followed by Cou (95.8 ± 4.2%), Ami (93.8 ± 3.6%), and then the control (89.4 ± 4.1%). When attempting to calculate LT values, the LT_5, LT_{10}, and LT_{30} values for all treatment groups could not be reliably estimated because survival rates remained above 90% (Table 3).

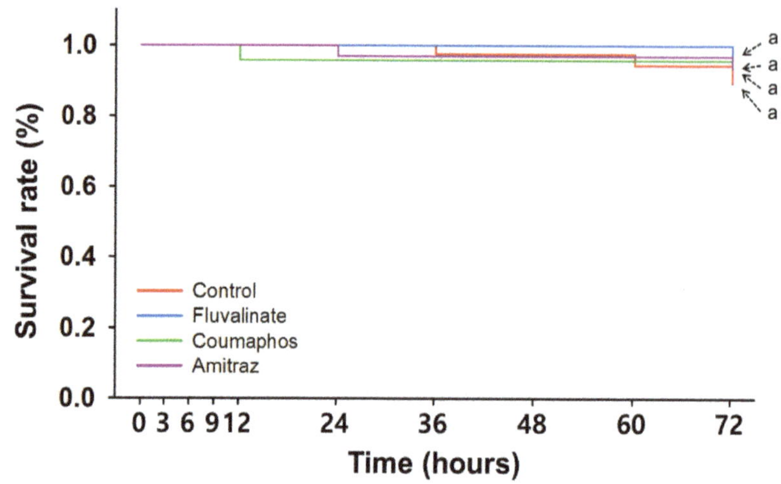

Figure 2. Time-dependent survival rate of nurse bees treated with synthetic miticides. Overall survival curve analyzed by Kaplan–Meier survival analysis with Log-rank test. Statistically significant differences ($p < 0.05$) among miticide treatments are indicated by different letters.

Table 2. Survival rates (arithmetic mean ± standard error) of nurse bees at multiple time points after single and consecutive treatments with miticides.

Time (hpt [a])	Control	Single Treatment Fluvalinate	Coumaphos	Amitraz	Oxalic Acid	Formic Acid	Consecutive Treatment Fluvalinate + Oxalic Acid	Coumaphos + Oxalic Acid	Amitraz + Oxalic Acid	Fluvalinate + Formic Acid	Coumaphos + Formic Acid	Amitraz + Formic Acid
3	100 ± 0.0	100 ± 0.0	100 ± 0.0	100 ± 0.0	100 ± 0.0	92.5 ± 5.0	100 ± 0.0	100 ± 0.0	100 ± 0.0	97.9 ± 2.1	94.2 ± 4.2	91.7 ± 5.3
6	100 ± 0.0	100 ± 0.0	100 ± 0.0	100 ± 0.0	100 ± 0.0	84.2 ± 7.3	100 ± 0.0	100 ± 0.0	100 ± 0.0	96.1 ± 2.5	92.1 ± 4.1	83.3 ± 7.7
9	100 ± 0.0	100 ± 0.0	100 ± 0.0	100 ± 0.0	100 ± 0.0	84.2 ± 7.3	100 ± 0.0	100 ± 0.0	100 ± 0.0	96.1 ± 2.5	92.1 ± 4.1	79.2 ± 9.5
12	100 ± 0.0	100 ± 0.0	95.8 ± 4.2	100 ± 0.0	100 ± 0.0	84.2 ± 7.3	100 ± 0.0	95.8 ± 4.2	91.7 ± 4.2	94.2 ± 2.6	82.9 ± 6.5	77.5 ± 8.9
24	100 ± 0.0	100 ± 0.0	95.8 ± 4.2	96.9 ± 3.1	87.5 ± 12.5	73.8 ± 8.9	83.3 ± 8.3	87.5 ± 7.2	83.3 ± 8.3	94.2 ± 2.6	75.4 ± 5.3 **	69.2 ± 5.8 ***
36	97.5 ± 2.5	100 ± 0.0	95.8 ± 4.2	96.9 ± 3.1	75 ± 12.5	70.6 ± 10.6 *	75.0 ± 7.2 *	79.2 ± 11.0	66.7 ± 4.2 ***	94.2 ± 2.6	73.8 ± 6.4 *	65.8 ± 6.6 **
48	97.5 ± 2.5	100 ± 0.0	95.8 ± 4.2	96.9 ± 3.1	68.8 ± 18.8	68.1 ± 8.4 *	70.8 ± 11.0 *	70.8 ± 4.2 ***	66.7 ± 4.2 ***	92.1 ± 2.5	71.7 ± 7.4 *	64.2 ± 7.5 **
60	94.4 ± 3.3	100 ± 0.0	95.8 ± 4.2	96.9 ± 3.1	62.5 ± 12.5 *	68.1 ± 8.4 *	70.8 ± 11.0	66.7 ± 4.2 ***	66.7 ± 4.2 ***	92.1 ± 2.5	71.7 ± 7.4 *	62.1 ± 7.9 *
72	89.4 ± 4.1	96.9 ± 3.1	95.8 ± 4.2	93.8 ± 3.6	62.5 ± 12.5	68.1 ± 8.4	70.8 ± 11.0	62.5 ± 7.2 *	66.7 ± 4.2 *	85.9 ± 3.9	69.6 ± 7.5	62.1 ± 7.9 *

[a] hours post-treatment (hpt); survival rates of each treatment group were statistically compared to the control group at the same time point. Statistical significance is denoted using asterisks: *, $p < 0.05$; **, $p < 0.01$; and ***, $p < 0.001$.

Taken together, when exposed to single synthetic miticides at field-realistic concentrations, while slight variations in survival rates were observed among treatment groups at each time point, the overall survival rates of nurse bees were not significantly different from those of nurse bees in the control group over the 72 h observation period.

3.2. Single Treatments with Organic Miticides

After single treatments with either OA or FA, the trends in nurse bee survival over time were significantly reduced when compared to the control group ($p < 0.05$); however, no statistical difference could be identified between the two organic miticide treatments ($p > 0.05$) (Table S2 and Figure 3). Notably, FA-treated nurse bees exhibited a more rapid decline in survival than those treated with OA within the 24 hpt (Figure 3).

Table 3. Estimated median lethal times (LTs) of the five single and six consecutive miticide treatments against nurse honey bees.

Treatment		Estimated Lethal Times (h)		
		LT_5	LT_{10}	LT_{30}
Single treatment	Fluvalinate	>72 *	>72 *	>72 *
	Coumaphos	>72 *	>72 *	>72 *
	Amitraz	>72 *	>72 *	>72 *
	Oxalic acid	16.04	22.46	44.09
	Formic acid	1.49	4.92	29.41
Consecutive treatment	Fluvalinate + Oxalic acid	12.62	18.67	38.99
	Coumaphos + Oxalic acid	13.18	19.98	42.74
	Amitraz + Oxalic acid	9.93	15.77	43.11
	Fluvalinate + Formic acid	15.71	70.54	>72 *
	Coumaphos + Formic acid	3.45	7.35	56.50
	Amitraz + Formic acid	1.15	3.74	24.04

* Honey bee groups treated with the miticide or miticide combination are estimated to reach the corresponding LT mortality level after 72 hpt.

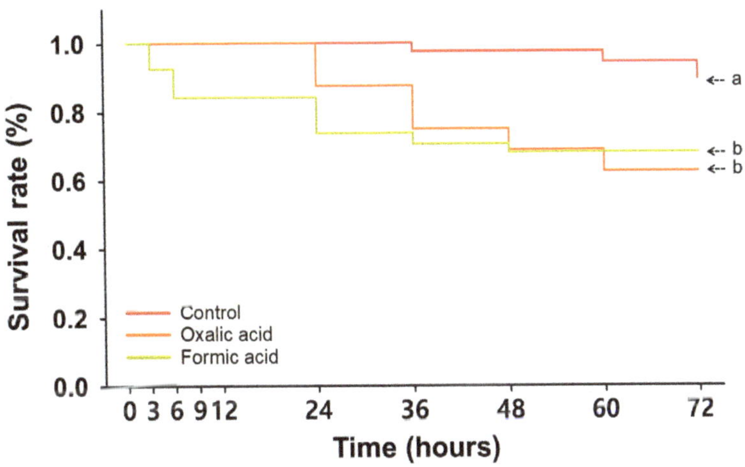

Figure 3. Time-dependent survival rate of nurse bees treated with organic miticides. Overall survival curve analyzed by Kaplan–Meier survival analysis with Log-rank test. Statistically significant differences ($p < 0.05$) among miticide treatments are indicated by different letters.

When comparing survival at each observation period (Table 2), the OA-treated group only differed significantly from the control group at 60 hpt. In contrast, the FA treatment significantly reduced survival in the period from 36 to 60 hpt (Table 2). At 24 h after the second treatment, both organic miticides slightly increased mortality compared to the control; however, the survival rates of the OA (87.5 ± 12.5%, $p > 0.05$) and FA (73.8 ± 8.9%, $p > 0.05$) groups were not statistically different from that of control group (Table 2). At 48 hpt, OA-treated nurse bees exhibited a slightly lower survival rate of 68.8 ± 18.8%, which did not significantly differ from that of nurse bees in the control group (97.5 ± 2.5%) ($p > 0.05$). However, the FA treatment significantly reduced survival rates to 68.1 ± 8.4% ($p < 0.05$) after 48 hpt. By 72 hpt, survival rates were highest in the control group (89.4 ± 4.1%), followed by FA (68.1 ± 8.4%) and then OA (62.5 ± 12.5%), with no statistically significant difference between the two organic miticide treatments and the control ($p > 0.05$).

Table 3 presents the LT values for the OA and FA single-treatment groups. For OA, the LT_5, LT_{10}, and LT_{30} values were 16.0, 22.5, and 44.1 h, respectively. In contrast, FA exhibited a faster toxic response, with LT_5, LT_{10}, and LT_{30} values of 1.5, 4.9, and 29.4 h, respectively, indicating more rapid toxicity than OA.

In summary, FA and OA treatments significantly decreased overall survival rates (Figure 3), although survival did not differ significantly from the control group at certain time points. Additionally, FA produced a more rapid onset of toxic effects in nurse bees than OA (Table 3).

3.3. Consecutive Treatments with Synthetic and Organic Miticides

To assess the toxicity of consecutive miticide treatments, nurse bees were exposed to synthetic and then organic miticides, with a 24 h interval between applications. Based on overall survival trends (Figure 4), nurse bees in the OA-combined treatment groups exhibited similar mortality patterns regardless of whether they had received a synthetic miticide treatment or the compound therein ($p > 0.05$) (Table S3 and Figure 4). However, all OA-treated groups experienced significantly higher mortality than the control group ($p < 0.05$) (Table S3).

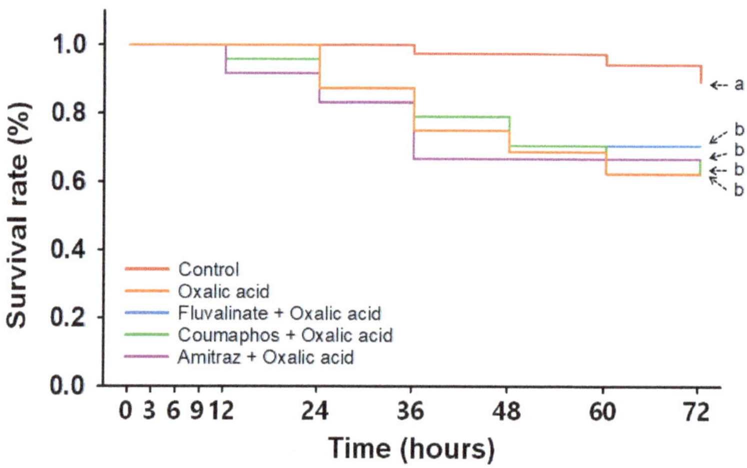

Figure 4. Time-dependent survival rate of nurse bees consecutively treated with synthetic miticides and oxalic acid. Overall survival curve analyzed by Kaplan–Meier survival analysis with Log-rank test. Statistically significant differences ($p < 0.05$) among miticide treatments are indicated by different letters.

When survival rates between the control group and OA-combined groups were compared at each time point (Table 2), treatment with Flu + OA produced nonsignificant differences across all time points except 36 and 48 hpt, whereas the Cou + OA and Ami + OA treatments induced significantly lower survivorship starting at 48 and 36 hpt, respectively. Specifically, at 24 hpt, the survival rate of the control group was 100.0 ± 0.0%, while it was slightly reduced in the Flu + OA (83.3 ± 8.3%, $p > 0.05$), Cou + OA (87.5 ± 7.2%, $p > 0.05$), and Ami + OA (83.3 ± 8.3%, $p > 0.05$) groups. In contrast, by 48 hpt, survival rates in all OA-combined groups were significantly lower than those of the control group (97.5 ± 2.5%), with survival rates recorded at 70.8 ± 11.0%, 70.8 ± 4.2%, and 66.7 ± 4.2% for the Flu + OA, Cou + OA, and Ami + OA groups ($p < 0.05$), respectively. By 72 hpt, survival rates in the Cou + OA and Ami + OA groups remained significantly lower than those in the control group ($p < 0.05$), but those in the Flu + OA and single OA treatment groups did not ($p > 0.05$). After 72 hpt, control nurse bees had the highest survival, (89.4 ± 4.1%), followed by those of the Flu + OA (70.8 ± 11.0%), Ami + OA (66.7 ± 4.2%), and then the Cou + OA (62.5 ± 7.2%) and OA-only (62.5 ± 12.5%) groups (Table 2).

In the consecutive treatments of synthetic miticides and FA (Figure 5 and Table S4), the overall survival trends of the Cou + FA and Ami + FA groups were significantly lower than that of control group ($p < 0.05$); however, the Flu + FA treatment did not produce a significantly different survival trend compared with the untreated control group ($p > 0.05$). Compared to the other treatments, the survival trend of nurse bees after the FA single treatment was statistically different from those of the control and Flu + FA groups ($p < 0.05$), but similar to those of the Cou + FA and Ami + FA groups ($p > 0.05$).

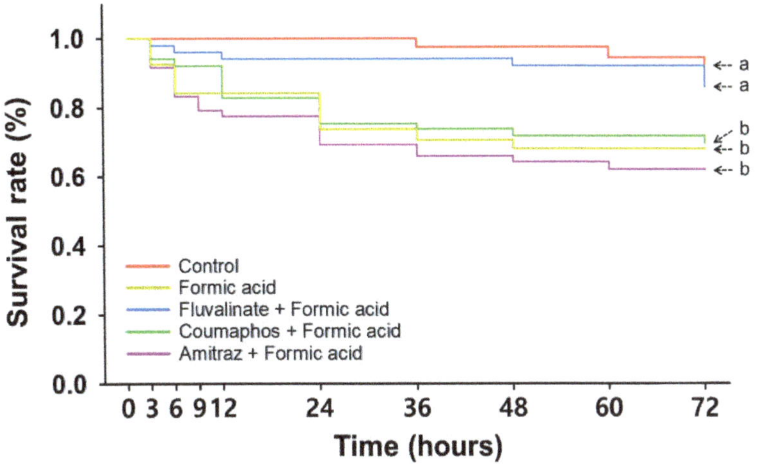

Figure 5. Time-dependent survival rate of nurse bees consecutively treated with synthetic miticides and formic acid. Overall survival curve analyzed by Kaplan–Meier survival analysis with Log-rank test. Statistically significant differences ($p < 0.05$) among miticide treatments are indicated by different letters.

In comparisons at different time points (Table 2), Flu + FA exhibited no significant difference in survival rates at any recorded time, while Ami + FA showed significantly lower survival at 24 hpt and thereafter. Furthermore, the Cou + FA and FA-only groups had significantly lower survival rates than the control group from 24 to 60 hpt and 36 to 60 hpt, respectively. At 24 hpt, survival in the Flu + FA group (94.2 ± 2.6%) exhibited a slight, nonsignificant decline ($p > 0.05$), whereas survival in the Cou + FA (75.4 ± 5.3%, $p < 0.05$) and Ami + FA (69.2 ± 5.8%, $p < 0.05$) groups was significantly lower than in the control group (100.0 ± 0.0%). Though survival in the FA single-treatment group (73.8 ± 8.9%) showed a clear reduction at 24 hpt, it was not significantly different from that in the control ($p > 0.05$). At 48 hpt, the Flu + FA group (92.1 ± 2.5%, $p > 0.05$) exhibited a survival rate similar to that of the control (97.5 ± 2.5%), but the Cou + FA (71.7 ± 7.4%) and Ami + FA (64.2 ± 7.5%) treatments significantly reduced survival ($p < 0.05$). In contrast to the 24 hpt results, the FA-only group exhibited significantly lower survival at 48 hpt (68.1 ± 8.4%, $p < 0.05$). By 72 hpt, survival rates among most OA-treated groups were not significantly different from that of the control group ($p > 0.05$ for Flu + FA, Cou + FA, and FA-only), but the Ami + FA group remained significantly lower ($p < 0.05$). At 72 hpt, the control group boasted the highest survival (89.4 ± 4.1%), followed by the Flu + FA (85.9 ± 3.9%), Cou + FA (69.6 ± 7.5%), and FA-only (68.1 ± 8.4%) groups, with the Ami + FA treatment resulting in the lowest (62.1 ± 7.9%).

When LT values of OA-combined groups were calculated using extrapolation (Table 3), LT_5 values of the Flu + OA, Cou + OA, and Ami + OA groups were 12.6, 13.2, and 9.9 h, respectively, and LT_{10} values were 18.7, 20.0, and 15.8 h, respectively. The LT_{30} values of nurse bees consecutively treated with synthetic miticides and OA were estimated to be between 39.0 and 43.1 h. When FA was combined with synthetic miticides, LT values varied greatly depending on the specific chemical used. The LT_5 and LT_{10} values of Flu + FA-treated nurse bees were calculated to be 15.7 h and 70.5 h, respectively, but the LT_{30} could not be calculated because survival rates remained above 90% after treatment. In contrast, the LT_5, LT_{10}, and LT_{30} values of the Cou + FA treatment were 3.5, 7.4, and 56.5 h, respectively, and those of the Ami + FA treatment were 1.2, 3.7, and 24.0 h, respectively.

In summary, after consecutive treatments with synthetic miticides and then organic miticides, survivorship of nurse bees was generally significantly lower than that of nurse bees in the control group, except for the combination of Flu and FA (Figure 5). Interestingly, similar survival trends were usually observed between each single organic miticide treatment and the consecutive treatments involving the same organic miticide, suggesting that organic miticides may be particularly toxic to nurse bees, but no synergistic effects were produced by synthetic miticides (Table 2).

4. Discussion

This study evaluated the individual and combined toxicities of synthetic (Flu, Cou, and Ami) and organic (FA and OA) miticides on nurse bees of *A. mellifera* using field-realistic concentrations and exposure protocols that reflect common apicultural practices. Our findings demonstrated that while synthetic miticides exerted minimal acute toxicity, organic miticides—particularly FA—caused substantial mortality. Furthermore, the toxicity observed in consecutive treatments was largely attributable to the organic miticides rather than any synergistic effects with synthetic agents.

Aligning with some studies [7,15], nurse bees exhibited high survival rates following single exposures to Flu, Cou, and Ami at field concentrations, with no significant differences from the control group (Table S1 and Figure 2). However, these results differ from other studies reporting higher toxicities at comparable or even lower doses [22,50,51]. Such discrepancies may be explained by methodological differences, including differences in the physiological age of the treated honey bees, application methods, and observation durations. For instance, the use of nurse bees—rather than foragers or mixed-age workers—may have influenced susceptibility, given the demonstrated age-dependent variation in detoxification enzyme expression and immune function in honey bees [52,53].

In contrast, both FA and OA significantly decreased survival (Table S2 and Figure 3), with FA inducing a more rapid onset. The estimated lethal times (LT_5, LT_{10}, and LT_{30}) for FA were markedly shorter than for OA, indicating a more acute toxic response (Table 3). These findings align with previous reports that high doses or the improper application of organic miticides can cause physiological stress, cellular damage, and behavioral abnormalities [26,27]. The topical application to the head, thorax, and abdomen used in our study likely facilitated rapid systemic absorption, potentially exacerbating neurotoxic effects. Notably, the application of acids to head tissues can induce local acidosis in neural tissues, leading to protein denaturation and neuronal damage [24,54].

In the sequential treatment assays, combined applications of synthetic and organic miticides generally did not produce effects beyond those observed with organic miticides alone, and thus, did not demonstrate additive or synergistic effects. Survival rates in most combined groups were not statistically different from the corresponding single organic miticide treatment (Figures 4 and 5 and Tables S3 and S4), suggesting that the organic agents were primarily responsible for the observed toxicity. This is further supported by the lack of significant differences in LT values between single and combined exposures involving organic miticides. An exception was the combination of Flu and FA. After treatment with this combination, nurse bees maintained survival rates comparable to the control group bees' survival rates throughout the observation period. This unexpected result may be due to the two compounds producing opposing physiological effects; Flu has been reported to increase respiratory rates [55,56], while FA is known to inhibit mitochondrial respiration [24].

The potential antagonism between these compounds warrants further investigation and highlights the importance of metabolic interactions between miticides. Interestingly, our findings contradict those of Johnson et al. [57], who reported reduced toxicity when combining Ami and OA, suggesting potential protective interactions. In our study, the Ami + OA treatment induced a relatively strong toxicity, much greater than that of Ami alone (Table 2). This discrepancy may arise from differences in treatment order; Johnson et al. [57] applied OA before applying Ami, which may have induced hydrolysis and the degradation of Ami [58], whereas our protocol applied Ami first. Thus, the persistence of Ami under acidic conditions may explain the enhanced toxicity in our sequential treatment.

The accumulation of miticide residues in hive matrices such as wax, pollen, and bee bread, as reported in previous studies [32,40], represents a persistent source of exposure for nurse bees. Residues of Flu, Cou, and Ami degradation products have been detected at considerable concentrations, sometimes exceeding 200 ppm. Given that nurse bees are in prolonged contact with hive materials, they are particularly vulnerable to chronic exposure and potential sublethal effects.

Taken together, our data demonstrate that organic miticides—particularly FA—pose a greater acute risk to nurse bees than their synthetic counterparts when applied at field-relevant concentrations. While synthetic miticides did not exacerbate organic miticide toxicity under the conditions tested, repeated or improper application in field settings may lead to cumulative effects. The variability in survival rates between treatment combinations underscores the important influences of application order, compound interactions, and the physiological state of the honey bees. Although our study focused on acute mortality, we recognize that sublethal effects of miticides—including those on glandular development, gut microbiota, and larval care behaviors—may also significantly impair nurse bee function and colony health. Future investigations incorporating physiological and behavioral endpoints are warranted to fully characterize miticide-related risks.

In conclusion, our findings underscore the necessity for caution when using organic miticides, especially in combination treatments. While synthetic miticides exhibited minimal acute toxicities at field concentrations, the inclusion of organic miticides significantly compromised nurse bee survival. Miticide application strategies should be reevaluated with the aim of minimizing collateral harm to critical in-hive populations such as nurse bees, whose loss may disrupt brood care and colony development. Future research should investigate the chronic and sublethal effects of miticide combinations, as well as their physiological mechanisms of action. In addition, the 72 h observational period employed in this study was selected to capture acute mortality trends based on previous miticide toxicity studies. However, we acknowledge that extended observation periods (e.g., $\geq$10 days) may reveal delayed or cumulative effects not detected within this timeframe. Further long-term studies are warranted to fully assess the chronic impacts of miticide exposure on nurse bee survival.

Supplementary Materials: The following supporting information can be downloaded at: https://www.mdpi.com/article/10.3390/insects16070657/s1, Table S1: *p* value from Log-Rank test of Kaplan-Meier survival analysis in single treatment groups with synthetic miticide.; Table S2: *p* value from Log-Rank test of Kaplan-Meier survival analysis in single treatment groups with organic miticide.; Table S3: *p* value from Log-Rank test of Kaplan-Meier survival analysis in consecutive treatment groups with synthetic miticide and oxalic acid.; Table S4: *p* value from Log-Rank test of Kaplan-Meier survival analysis in consecutive treatment groups with synthetic miticide and formic acid.

Author Contributions: Conceptualization, H.K., S.H.L. and Y.H.K.; methodology, H.K.; validation, H.K., E.Y. and J.C.; investigation, H.K., E.Y. and J.C.; resources, Y.H.K.; writing—original draft, H.K., E.Y. and Y.H.K.; writing—review and editing, S.H.L. and Y.H.K.; project administration, Y.H.K.; funding acquisition, Y.H.K. All authors have read and agreed to the published version of the manuscript.

Funding: This study was supported by a research fund (RS-2025-02303308) from the Rural Development Administration, Republic of Korea.

Data Availability Statement: The original contributions presented in this study are included in the article/Supplementary Materials. Further inquiries can be directed to the corresponding author.

Conflicts of Interest: The authors declared no conflicts of interest.

References

1. Bruckner, S.; Wilson, M.; Aurell, D.; Rennich, K.; Vanengelsdorp, D.; Steinhauer, N.; Williams, G.R. A national survey of managed honey bee colony losses in the USA: Results from the Bee Informed Partnership for 2017–18, 2018–2019, and 2019–20. *J. Apic. Res.* **2023**, *62*, 429–443. [CrossRef]
2. Gray, A.; Adjlane, N.; Arab, A.; Ballis, A.; Brusbardis, V.; Bugeja Douglas, A.; Cadahía, L.; Charrière, J.D.; Chlebo, R.; Coffey, M.F. Honey bee colony loss rates in 37 countries using the COLOSS survey for winter 2019–2020: The combined effects of operation size, migration and queen replacement. *J. Apic. Res.* **2023**, *62*, 204–210. [CrossRef]
3. Kim, H.K. The Effect of Honey Bee Mites on the Winter Colony Losses. *J. Apic.* **2022**, *37*, 291–299.
4. Dainat, B.; Evans, J.D.; Chen, Y.P.; Gauthier, L.; Neumann, P. Predictive markers of honey bee colony collapse. *PLoS ONE* **2012**, *7*, e32151. [CrossRef]
5. Kim, Y.H.; Lee, S.K. Current status of fluvalinate resistance in *Varroa destructor* in Korea and suggestion for possible solution. *J. Apic.* **2022**, *37*, 301–313.
6. Pervez, M.; Manzoor, F.; Farooq, U. Differential expression of antimicrobial peptide encoding genes of honey bee (*Apis mellifera*) after Varroa mite (*Varroa destructor*) infestation. *J. Apic. Res.* **2024**, *64*, 871–878. [CrossRef]
7. Rademacher, E.; Harz, M.; Schneider, S. Effects of oxalic acid on *Apis mellifera* (Hymenoptera: Apidae). *Insects* **2017**, *8*, 84. [CrossRef]
8. APHIS. *Varroosis of Honey Bees*; United States Department of Agriculture (USDA): Riverdale, MD, USA, 2023; pp. 1–2.
9. Rosenkranz, P.; Aumeier, P.; Ziegelmann, B. Biology and control of *Varroa destructor*. *J. Invertebr. Pathol.* **2010**, *103*, S96–S119. [CrossRef]
10. Ostiguy, N.; Drummond, F.A.; Aronstein, K.; Eitzer, B.; Ellis, J.D.; Spivak, M.; Sheppard, W.S. Honey bee exposure to pesticides: A four-year nationwide study. *Insects* **2019**, *10*, 13. [CrossRef]
11. Perez-Cobo, I.; Fernández-Alba, A.R.; Hernando, M.D. First national survey of residues of active substances in honeybee apiaries across Spain between 2012 and 2016. *Sci. Total Environ.* **2022**, *838*, 155614. [CrossRef]
12. Guo, L.; Fan, X.y.; Qiao, X.; Montell, C.; Huang, J. An octopamine receptor confers selective toxicity of amitraz on honeybees and Varroa mites. *Elife* **2021**, *10*, e68268. [CrossRef] [PubMed]
13. Mao, W.; Schuler, M.A.; Berenbaum, M.R. CYP9Q-mediated detoxification of acaricides in the honey bee (*Apis mellifera*). *Proc. Natl. Acad. Sci. USA* **2011**, *108*, 12657–12662. [CrossRef] [PubMed]
14. Liesivuori, J.; Savolainen, A.H. Methanol and formic acid toxicity: Biochemical mechanisms. *Pharmacol. Toxicol.* **1991**, *69*, 157–163. [CrossRef]
15. Johnson, R.M.; Ellis, M.D.; Mullin, C.A.; Frazier, M. Pesticides and honey bee toxicity–USA. *Apidologie* **2010**, *41*, 312–331. [CrossRef]
16. Locke, B.; Forsgren, E.; Fries, I.; de Miranda, J.R. Acaricide treatment affects viral dynamics in *Varroa destructor*-infested honey bee colonies via both host physiology and mite control. *Appl. Environ. Microbiol.* **2012**, *78*, 227–235. [CrossRef]
17. Frost, E.H.; Shutler, D.; Hillier, N.K. Effects of fluvalinate on honey bee learning, memory, responsiveness to sucrose, and survival. *J. Exp. Biol.* **2013**, *216*, 2931–2938. [CrossRef]
18. Ilyasov, R.; Lim, S.; Lee, M.L.; Kwon, H.W.; Nikolenko, A. Effect of miticides amitraz and fluvalinate on reproduction and productivity of honey bee *Apis mellifera*. *Uludag Bee J.* **2021**, *21*, 21–30. [CrossRef]
19. Sokół, R. Effects of long-term persistence of Fluwarol [fluvalinate] on honey bee colonies. *Med. Weter* **1996**, *52*, 718–720.
20. Haarmann, T.; Spivak, M.; Weaver, D.; Weaver, B.; Glenn, T. Effects of fluvalinate and coumaphos on queen honey bees (Hymenoptera: Apidae) in two commercial queen rearing operations. *J. Econ. Entomol.* **2002**, *95*, 28–35. [CrossRef]

21. Bevk, D.; Kralj, J.; Čokl, A. Coumaphos affects food transfer between workers of honeybee *Apis mellifera*. *Apidologie* **2012**, *43*, 465–470. [CrossRef]
22. Gregorc, A.; Bowen, I.D. Histochemical characterization of cell death in honeybee larvae midgut after treatment with Paenibacillus larvae, amitraz and oxytetracycline. *Cell Biol. Int.* **2000**, *24*, 319–324. [CrossRef] [PubMed]
23. Strachecka, A.; Paleolog, J.; Olszewski, K.; Borsuk, G. Influence of amitraz and oxalic acid on the cuticle proteolytic system of *Apis mellifera* L. workers. *Insects* **2012**, *3*, 821–832. [CrossRef] [PubMed]
24. Bolli, H.; Bogdanov, S.; Imdorf, A.; Fluri, P. Action of formic acid on *Varroa jacobsoni* Oud and the honeybee (*Apis mellifera* L). *Apidologie* **1993**, *24*, 51–57. [CrossRef]
25. Westcott, L.C.; Winston, M.L. Chemical acaricides in *Apis mellifera* (Hymenoptera: Apidae) colonies; do they cause nonlethal effects? *Can. Entomol.* **1999**, *131*, 363–371. [CrossRef]
26. Gregorc, A.; Pogacnik, A.; Bowen, I.D. Cell death in honeybee (*Apis mellifera*) larvae treated with oxalic or formic acid. *Apidologie* **2004**, *35*, 453–460. [CrossRef]
27. Underwood, R.M.; Currie, R.W. Effects of release pattern and room ventilation on survival of varroa mites and queens during indoor winter fumigation of honey bee colonies with formic acid. *Can. Entomol.* **2007**, *139*, 881–893. [CrossRef]
28. Gregorc, A.; Škerl, M.I.S. Toxicological and immunohistochemical testing of honeybees after oxalic acid and rotenone treatments. *Apidologie* **2007**, *38*, 296–305. [CrossRef]
29. Maggi, M.D.; Ruffinengo, S.R.; Damiani, N.; Sardella, N.H.; Eguaras, M.J. First detection of *Varroa destructor* resistance to coumaphos in Argentina. *Exp. Appl. Acarol.* **2009**, *47*, 317–320. [CrossRef]
30. Maggi, M.D.; Ruffinengo, S.R.; Negri, P.; Eguaras, M.J. Resistance phenomena to amitraz from populations of the ectoparasitic mite *Varroa destructor* of Argentina. *Parasitol. Res.* **2010**, *107*, 1189–1192. [CrossRef]
31. Lee, J.; Moon, K.; Cho, S.; Lim, Y.; Kim, S.; Kim, S.-b.; Han, S.-M.; Kim, Y.H.; Lee, S.H. Establishment and application of bioassay-and molecular marker-based methods for monitoring fluvalinate resistance of Varroa mites. *Pest. Biochem. Physiol.* **2023**, *197*, 105655. [CrossRef]
32. Mullin, C.A.; Frazier, M.; Frazier, J.L.; Ashcraft, S.; Simonds, R.; VanEngelsdorp, D.; Pettis, J.S. High levels of miticides and agrochemicals in North American apiaries: Implications for honey bee health. *PLoS ONE* **2010**, *5*, e9754. [CrossRef] [PubMed]
33. Kast, C.; Kilchenmann, V.; Droz, B. Distribution of coumaphos in beeswax after treatment of honeybee colonies with CheckMite® against the parasitical mite *Varroa destructor*. *Apidologie* **2020**, *51*, 112–122. [CrossRef]
34. Premrov Bajuk, B.; Babnik, K.; Snoj, T.; Milčinski, L.; Pislak Ocepek, M.; Škof, M.; Jenčič, V.; Filazi, A.; Štajnbaher, D.; Kobal, S. Coumaphos residues in honey, bee brood, and beeswax after Varroa treatment. *Apidologie* **2017**, *48*, 588–598. [CrossRef]
35. Morales, M.M.; Ramos, M.J.G.; Vázquez, P.P.; Galiano, F.J.D.; Valverde, M.G.; López, V.G.; Flores, J.M.; Fernández Alba, A.R. Distribution of chemical residues in the beehive compartments and their transfer to the honeybee brood. *Sci. Total Environ.* **2020**, *710*, 136288. [CrossRef]
36. Murcia-Morales, M.; Díaz-Galiano, F.J.; Guitérrez-Tirado, I.; Flores, J.M.; Van der Steen, J.J.; Fernández-Alba, A.R. Dissipation and cross-contamination of miticides in apiculture. Evaluation by APIStrip-based sampling. *Chemosphere* **2021**, *280*, 130783. [CrossRef]
37. Jiménez, J.J.; Bernal, J.L.; del Nozal, M.J.; Martín, M.T. Residues of organic contaminants in beeswax. *Eur. J. Lipid Sci. Techno.* **2005**, *107*, 896–902. [CrossRef]
38. Jeon, S. Relation Between the Honey Bee Mortality and the Pesticide Residue Detected During the Pear and Apple Blooming Season. Ph.D. Thesis, Andong National University, National University of Andong Graduate School, Andong-si, Republic of Korea, 2017.
39. Shin, Y. Simultaneous Analysis of Pesticide Multiresidues in Human Serum, Urine, Apiculture Samples, and Representative Crops Using Tandem Mass Spectrometry. Ph.D. Thesis, Seoul National University, Seoul National University Graduate School, Seoul, Republic of Korea, 2018.
40. Calatayud-Vernich, P.; Calatayud, F.; Simó, E.; Aguilar, J.A.P.; Picó, Y. A two-year monitoring of pesticide hazard in-hive: High honey bee mortality rates during insecticide poisoning episodes in apiaries located near agricultural settings. *Chemosphere* **2019**, *232*, 471–480. [CrossRef]
41. Jeong, S.; Lee, C.; Kim, D.; Jung, C. Questionnaire study on the overwintering success and pest management of honeybee and damage assessment of Vespa hornets in Korea. *Korean J. Apic.* **2016**, *31*, 201–210. [CrossRef]
42. Johnson, B.R. Division of labor in honeybees: Form, function, and proximate mechanisms. *Behav. Ecol. Sociobiol.* **2010**, *64*, 305–316. [CrossRef]
43. Fakhimzadeh, K. Potential of super-fine ground, plain white sugar dusting as an ecological tool for the control of varroasis in the honey bee (*Apis mellifera*). *Am. Bee J.* **2000**, *140*, 487–491.
44. Bogdanov, S.; Kilchenmann, V.; Imdorf, A. Acaricide residues in some bee products. *J. Apic. Res.* **1998**, *37*, 57–67. [CrossRef]
45. Tremolada, P.; Bernardinelli, I.; Colombo, M.; Spreafico, M.; Vighi, M. Coumaphos distribution in the hive ecosystem: Case study for modeling applications. *Ecotoxicology* **2004**, *13*, 589–601. [CrossRef] [PubMed]

46. Johnson, R.M.; Pollock, H.S.; Berenbaum, M.R. Synergistic interactions between in-hive miticides in *Apis mellifera*. *J. Econ. Entomol.* **2009**, *102*, 474–479. [CrossRef]
47. Oh, D.; Kim, K.W.; Kim, K.M.; Akongte, P.N.; Kim, J.K.; Lee, C.; Park, B.S.; Kim, S.B.; Jo, Y.Y.; Choi, Y.S.; et al. Evaluation of vaporization amount of formic acid in different conditions for controlling honeybee mites. *J. Apic.* **2023**, *38*, 11–20.
48. Poquet, Y.; Bodin, L.; Tchamitchian, M.; Fusellier, M.; Giroud, B.; Lafay, F.; Buleté, A.; Tchamitchian, S.; Cousin, M.; Pelissier, M. A pragmatic approach to assess the exposure of the honey bee (*Apis mellifera*) when subjected to pesticide spray. *PLoS ONE* **2014**, *9*, e113728. [CrossRef]
49. Mesa, M.G.; Weiland, L.K.; Maule, A.G. Progression and severity of gas bubble trauma in juvenile salmonids. *Trans. Am. Fish. Soc.* **2000**, *129*, 174–185. [CrossRef]
50. Hillier, N.K.; Frost, E.H.; Shutler, D. Fate of dermally applied miticides fluvalinate and amitraz within honey bee (Hymenoptera: Apidae) bodies. *J. Econ. Entomol.* **2013**, *106*, 558–565. [CrossRef]
51. Santiago, G.P.; Otero-Colina, G.; Sánchez, D.M.; Guzmán, M.E.R.; Vandame, R. Comparing effects of three acaricides on *Varroa jacobsoni* (Acari: Varroidae) and *Apis mellifera* (Hymenoptera: Apidae) using two application techniques. *Fla. Entomol.* **2000**, *83*, 468–476. [CrossRef]
52. Nielsen, S.A.; Brødsgaard, C.J.; Hansen, H. Effects on detoxification enzymes in different life stages of honey bees (*Apis mellifera* L., Hymenoptera: Apidae) treated with a synthetic pyrethroid (flumethrin). *Altern. Lab. Anim.* **2000**, *28*, 437–443. [CrossRef]
53. Vannette, R.L.; Mohamed, A.; Johnson, B.R. Forager bees (*Apis mellifera*) highly express immune and detoxification genes in tissues associated with nectar processing. *Sci. Rep.* **2015**, *5*, 16224. [CrossRef]
54. Kraig, R.P.; Wagner, R.J. Acid-induced changes of brain protein buffering. *Brain Res.* **1987**, *410*, 390–394. [CrossRef] [PubMed]
55. Mckee, M.J.; Knowles, C.O. Effects of pyrethroids on respiration in the twospotted spider mite (Acari: Tetranychidae). *J. Econ. Entomol.* **1984**, *77*, 1376–1380. [CrossRef]
56. Kankaya, E. Susceptibility of fluvalinate and esfenvalerate on adult tarek *Alburnus tarichi* (Güldenstädt 1814). *Croat. J. Fish.* **2023**, *81*, 49–53. [CrossRef]
57. Johnson, R.M.; Dahlgren, L.; Siegfried, B.D.; Ellis, M.D. Acaricide, fungicide and drug interactions in honey bees (*Apis mellifera*). *PLoS ONE* **2013**, *8*, e54092. [CrossRef] [PubMed]
58. Corta, E.; Bakkali, A.; Berrueta, L.; Gallo, B.; Vicente, F. Kinetics and mechanism of amitraz hydrolysis in aqueous media by HPLC and GC-MS. *Talanta* **1999**, *48*, 189–199. [CrossRef]

insects

MDPI

Communication

A Novel Biosensor for the Early Detection of *Aethina tumida* via *Kodamaea ohmeri* in Honeybee Colonies

Paola Ghisellini [1], Patrizia Garbati [1], Marco Pietropaoli [2], Antonella Cersini [2], Gabriele Pietrella [2], Cristina Rando [1], Luca Giacomelli [1], Stefano Ottoboni [1], Giovanni Formato [2] and Roberto Eggenhöffner [1,*]

1 Department of Surgical Sciences and Integrated Diagnostics (DISC), University of Genova, Corso Europa 30, 16132 Genova, Italy; paola.ghisellini@unige.it (P.G.); p.r.garbati@gmail.com (P.G.); cristina.rando@unige.it (C.R.); luca.giacomelli@polistudium.it (L.G.); stefano.ottoboni@unige.it (S.O.)

2 Istituto Zooprofilattico Sperimentale del Lazio e della Toscana "M. Aleandri", Via Appia Nuova 1411, 00178 Rome, Italy; marco.pietropaoli@izslt.it (M.P.); antonella.cersini@izslt.it (A.C.); gabriele.pietrella@izslt.it (G.P.); giovanni.formato@izslt.it (G.F.)

* Correspondence: roberto.eggenhoffner@unige.it; Tel.: +39-01035338212

Simple Summary: This work presents a biosensor for the early detection of the small hive beetle *Aethina tumida* in honeybee colonies. The biosensor targets the symbiotic yeast *Kodamaea ohmeri*, associated with the beetle, which has been infesting hives and killing *Apis mellifera*. Based on QCM technology, this sensor detects, in honey samples, a peptide from *Kodamaea ohmeri*. It is a highly specific and sensitive device suitable for field tests to identify the infestation. Further studies are foreseen to improve the thresholds and robustness of this detection in an effort to protect sustainable agriculture and biodiversity.

Abstract: *Aethina tumida*, commonly known as the small hive beetle, poses a threat to honeybee populations, particularly *Apis mellifera*, across several European regions. Originating in sub-Saharan Africa, there is a risk of the infestation spreading from Calabria, a region in the south of Italy. The essential role played by *Apis* spp. in pollination biodiversity preservation, agricultural productivity, and the overall economy is related to the dangers posed by the invasion of *Aethina tumida*. Current detection methods often fail to identify infestations early, leading to significant colony losses. We focused on creating a biosensor to improve the detection of *Kodamaea ohmeri*, a recently identified yeast that coexists symbiotically with *Aethina tumida*. The biosensor was designed to exploit the highly sensitive quartz crystal microbalance to identify a specific peptide linked to *Kodamaea ohmeri* in honey specimens. Its gold-plated surface over quartz was functionalized with an antibody effective in recognizing the peptide associated with *Kodamaea ohmeri*, a potential warning for detecting *Aethina tumida*. Preliminary results support the possibility of using such biosensor technology to detect infestation and enhance colony management techniques for honeybees, enabling beekeepers to implement prompt and focused treatments. In addition, reducing the size and cost of these biosensors and offering user training would be very helpful in having them used in beekeeping.

Keywords: *Aethina tumida* detection; biosensors; quartz crystal microbalance; *Apis mellifera* health; *Kodamaea ohmeri* detection

1. Introduction

The protection of *Apis mellifera* populations is a priority, since they are highly instrumental in maintaining ecological balance and agricultural yield. These very important pollinators are, at the same time, facing serious threats in different regions [1], one of which is the Italian region of Calabria, mainly due to the infestation by *Aethina tumida*, also known as the small hive beetle (SHB) [2]. Since European honeybees do not have evolved defense mechanisms against SHB, like those from Africa, this pest is recognized as a serious threat to the viability of honeybee colonies [3]. One of the critical factors exacerbating this threat is the interaction between SHB and honeybee alarm pheromones within European honeybee colonies [4]. In addition to helping honeybees defend themselves, these pheromones attract beetles and facilitate their infestation. This situation is further aggravated by the presence of *Kodamaea ohmeri* (Ko) yeast, which is vectored by SHB and produces volatile compounds similar to alarm pheromones when grown on pollen within the hives. This not only attracts more beetles but also significantly disrupts the dynamics of the hive, leading to stress among the honeybees and, ultimately, their displacement from the hive [5]. Given the complex interplay of biological factors, managing the threat posed by SHB requires a multifaceted approach [6]. A critical challenge in this effort is the lack of efficient early detection tools. Traditional approaches using visual inspections or molecular methods are normally either insensitive, too costly, or cannot be employed in the field. Although molecular techniques are accurate, they are labor-intensive and usually unsuitable for routine monitoring by beekeepers. The proposed biosensor represents a novel approach, coupling quartz crystal microbalance technology with peptide-based specificity for Ko, a reliable biomarker for SHB infestations. This advancement addresses the limitations of existing tools, offering a scalable solution for early detection and potentially transforming beekeeping practices to improve resilience against colony losses.

However, effective management strategies also require a deep understanding of the biology of the beetle, particularly its lifecycle within the hive environment, where females lay eggs that give rise to larvae feeding on honey, pollen, and bee broods. This feeding behavior not only depletes the hive's resources but also contributes to honey spoilage through fermentation, further worsened by the structural damage to the comb caused by larval feeding activities. Moreover, the intricate association of SHB with Ko yeast adds another layer of complexity to the infestation dynamics. This yeast plays a crucial role in the fermentation process of honey within infested hives and has been detected in various developmental stages of SHB, indicating a symbiotic relationship that influences the pest's lifecycle and infestation patterns.

The detection of Ko yeast in association with SHB infestations [7] can be achieved through molecular techniques such as the PCR amplification of the ribosomal RNA gene (*26S rRNA*) for Ko and real-time PCR amplification of cytochrome oxidase for SHB. Reliable methods for isolating and identifying Ko from honey and hive samples were established in prior studies and protocols. The development of biosensors that use Ko's presence as an early indicator of SHB infestations has been made possible largely due to these molecular discoveries [8]. The purpose of this work is to show that a biosensor based on quartz microbalance (QCM) can detect Ko, which may be used as a sign of an infestation of SHB. In order to evaluate its utility as an early detection tool, further verification is necessary. To successfully protect honeybee health, ecosystems, and agricultural systems, such an approach should incorporate biological insights, mechanical control methods, chemical treatments, and regulatory laws [9].

2. Materials and Methods

2.1. Sample Collection, DNA Extraction, and PCR Amplification of Kodamaea ohmeri and RT-PCR of Aethina tumida

An on-site inspection was conducted following a report of infestation from a beekeeper, during which visual evidence of SHB presence was obtained as shown in Figure 1

Figure 1. Sample collection (**upper panel**) and details of *Aethina tumida* in the (**lower panel**).

The honey samples were collected from hives located in the red zone of Reggio Calabria, Italy. Honey samples were collected from infested and non-infested hives, and the DNA was extracted following the standard protocol reported [10] and using the QIAamp DNA Blood Mini Kit (QIAGEN, Hilden, Germany). Finally, it was diluted in 100 µL of a 10 mM Tris-Cl, pH 8,5 buffer, and stored at −20 °C until further use. Several tests were conducted to validate the presence of Ko and SHB in different hive components such as honey debris and comb for the subsequent molecular analyses, following the above protocols.

The PCR amplification of Ko was performed using the specific primers reported in Table 1 for the *26S rRNA* gene region of Ko [11], targeting a 320 bp fragment. The amplification conditions included the following: denaturation at 95 °C for 10 min, 40 cycles made up of 94 °C for 30 s, annealing at 53 °C for 30 s and 72 °C for 1 min, and elongation at 72 °C for 1 min. PCR products were verified via gel electrophoresis and visualized under UV light.

For SHB detection, the primer pair and probe for real-time PCR were selected from the literature and reported in Table 1. They are specific to SHB, since the comparison with the 15 sequences coding for Cytochrome Oxidase I (COI) available in GenBank. It was verified in silico that these sequences do not share homology with the COI of *Apis mellifera* L., the COI of the wax moth (*Galleria mellonella*) commonly found in apiaries, the COI of *Varroa destructor* (Acari: Varroidae), and the COI of *Tropilaelaps* spp. (Acari: Laelapidae). These primers and probe amplify a 109 bp fragment under the amplification conditions reported in the literature [12].

Table 1. Sequences of the primers for the PCR of *Kodamaea ohmeri* and the specific primers and probe for RT-PCR of *Aethina tumida*.

Ko Primers for PCR	Sequence
Ko Fw	5′-TAATTTGAAGATTGCGTCTTG-3′
Ko Rv	5′-TACCCACACTGACAATCTGAC-3′
SHB Primer/Probe name for RT-PCR	**Sequence**
Fw SBH207F	5′-TCTAAATACTACTTTCTTCGACCCAT(A/G)-3′
Rv SBH315R 5′	5′-TCCTGGTAGAATTAAAATATAAACTTCTGG-3′
SBH245T PROBE	5′-FAM-ATCCAATCCTATACCAACACTTATTTTGATTCTTCGGAC-3′-TAMRA

2.2. Identification of a Peptide from Kodamaea ohmeri

A specific peptide indicative of the presence of Ko was identified via the PCR analysis of ribosomal RNA sequences from contaminated honey [10]. In detail, the peptide sequence was obtained from the sequencing of the 300 bp amplicon related to the ITS2 of Ko isolated in 2015 from the SHB outbreak in the province of Reggio Calabria, Italy.

The nucleotide sequence of peptide was converted to a peptide sequence using the program In-Silicon Online (http://insilico.net/tools/biology/sequence_conversion, accessed on 25 March 2025). This peptide was synthesized by SynPeptide Co., Ltd. (Shanghai, China) and used to produce specific polyclonal antibodies (antipeptides) with Twin Helix (RHO-Milan, Italy).

2.3. Development of a Micro/Nanogravimetric Biosensor for Detecting Kodamaea ohmeri Contamination

A QCM Wi2 instrument from Novaetech S.R.L. (Pompei, Napoli, Italy) was selected for its compactness, versatility, and suitability for on-field applications. It is designed to accommodate gold-plated electrodes with a front electrode with a diameter of 11.20 mm, a working frequency of 5 MHz, and a sensitivity of 0.017 μg/Hz. This speaks of the QCM's ability to identify biomolecules since it can cover a wide range of sensitivities, which is essential for the precise measurements needed in advanced biosensing applications. The theoretical foundation of the operation of the QCM is grounded in the Sauerbrey equation, which correlates mass changes to frequency shifts, assuming that the deposited mass is rigidly anchored and evenly distributed across the electrode surface.

$$\Delta f = -\frac{2f_0^2}{A\sqrt{\rho_q \mu_q}}\Delta m \quad (1)$$

where f_0 is the resonance frequency (5 MHz in our case), Δf_0 is the frequency change (Hz), Δm is the mass change (g), A is the active crystal area (i.e., the deposition electrode area, cm^2), ρ_q = 2.648 g/cm^3 is the quartz density, and μ_q = 2.947 × 10^{11} g·cm^{-1}·s^{-2} (i.e., poise) is the AT-cut quartz shear modulus.

2.4. Surface Immobilization Protocol

A His-tagged protein A mutant derived from *Staphylococcus aureus* featured a 6xHis tag at its N-terminus and 5 fragment crystallizable (Fc)-specific binding sites, while nonspecific adsorption sites were deleted [13]. This protein, known for its high binding affinity for Fc fragments and expressed in the bacterial outer membrane, was supplied by BioVision Abcam, Cambridge, CB2 0AX, UK. A protein solution at a concentration of 1 mg/mL was prepared in 150 mM phosphate-buffered saline (PBS) by Sigma–Aldrich/Merck Life

Science S.r.L., Milan, Italy. A 200 µL aliquot of protein solution was applied to the gold QCM electrode. The high affinity of the 6xHis tag for gold facilitated the formation of a monolayer on the sensor surface, ensuring the precise molecular orientation of the protein. Following a two-hour incubation at room temperature to allow protein adhesion, the sample was rinsed with distilled water to remove any unbound material, followed by air-drying. Using a protocol akin to that used for protein A deposition, an antipeptide originating from a 1.2 mg/mL stock solution in 5 mM PBS was applied. We conducted frequency measurements using the QCM to verify the successful coating of the gold surface of the quartz crystal with a stable layer of protein A linked to the antipeptide [14], as discussed in Section 3.3.

2.5. Peptide Detection

For each assessment, 200 µL of a 0.125 mg/mL peptide solution in 5 mM PBS was deposited onto the functionalized quartz crystals. The choice of this dosage was made considering earlier research that found comparable peptide levels in hives afflicted with SHB. To properly evaluate the biosensor's sensitivity, further research will examine how it reacts to a larger range of concentrations.

These samples were then incubated at 4 °C overnight. The solution on the quartz surface was subsequently removed by allowing it to slide off. Washing procedures, as described above, were performed, and the quartz crystals were allowed to air-dry at room temperature. We used a control solution consisting only of PBS at the same concentration (5 mM) as the previous solution.

3. Results and Discussions

3.1. Detection of *Kodamaea ohmeri* and *Aethina tumida* in Hive Samples

Quantitative real-time PCR was used to detect SHB, while qualitative PCR was employed for Ko. A total of 15 samples were collected from hives, including honeycomb and hive bottom, in the SHB outbreak zone in Reggio Calabria, Italy, tested for the presence of SHB and Ko as reported in Table 2. The 15 samples tested positive for both SHB and Ko, excluding 2 RT-PCR invalid findings (sample ID 10 and 14). This supports the hypothesis that Ko could serve as an indicator of hive health, strongly indicating its specificity as a specific marker for SHB infestations.

Table 2. Real-time PCR detection of *Aethina tumida* and PCR of *Kodamaea ohmeri* in 15 samples (all honey samples, except 13, which was a honeycomb sample).

Sample ID	Real-Time PCR for *Aethina tumida* (Ct Value)	Qualitative PCR for *Kodamaea ohmeri* (Positive/Negative)
1	31.23	Positive
2	32.01	Positive
3	31.55	Positive
4	30.47	Positive
5	29	Positive
6	31.4	Positive
7	30.72	Positive
8	31.95	Positive
9	31.22	Positive

Table 2. *Cont.*

Sample ID	Real-Time PCR for *Aethina tumida* (Ct Value)	Qualitative PCR for *Kodamaea ohmeri* (Positive/Negative)
10	0	Positive
11	35	Positive
12	33.92	Positive
13	32.49	Positive
14	0	Positive
15	27.86	Positive

3.2. Molecular Detection of *Kodamaea ohmeri* in Hive Samples

In Figure 2, the gel electrophoresis results show the successful PCR amplification of the internal transcribed spacer (ITS2) region of Ko from hive swab samples. Positive PCR results are particularly visible in lanes 2, 3, 5, 6, 7, and 9, showing clear bands around 300 bp, confirming the presence of Ko, as reported in Table 2.

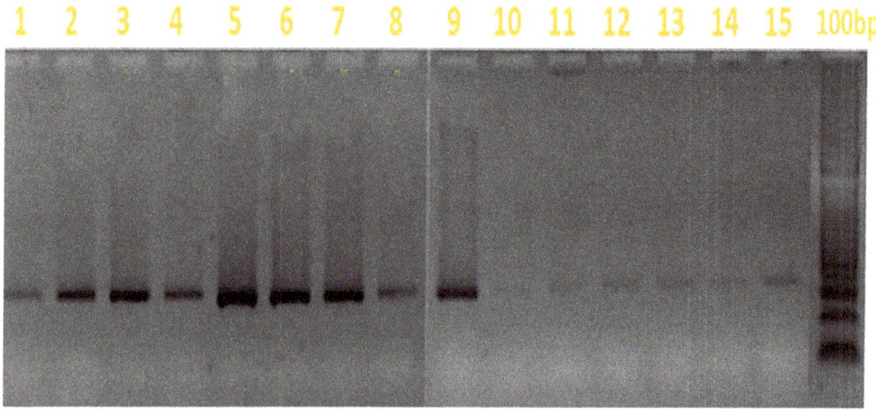

Figure 2. Gel electrophoresis of PCR products for *Kodamaea ohmeri*. Gel electrophoresis showing the PCR amplification of the ITS2 region of *Aethina tumida* infestation. Lanes 1–15 represent different hive samples, as follows: all samples were tested positive, samples 1 to 9 were taken from honeycomb swabs (the same samples as in Table 3) and positivity is particularly evident; samples 10 to 15 were also positive but less visible, and these were collected from hive bottom swabs. The molecular weight markers 100 bp plus the DNA ladder on the right indicate fragment sizes of 1500–100 bp.

Sequencing analysis of PCR products in Table 3 yielded a 97–99% match to the *Ko 26S rRNA* gene (Accession Number MG367286.1). The sequence was detected to correspond to the KBPAP56 internal transcribed spacer 2 of the Ko strain KBPAP56 and to a partial sequence of the ribosomal RNA gene subunit.

Table 3. Sequencing results in honeycomb swab samples testing positive for *Kodamaea ohmeri*.

Sample ID	Query Cover (%)	Identity (%)
1	99%	97%
2	96%	99%
3	96%	99%
4	97%	99%
5	99%	98%
7	96%	99%
8	94%	99%
9	97%	99%

3.3. Peptide and Anti-Peptide Productions

The above nucleotide sequence of MG367286.1 was converted to the following peptide sequence: ESRAARPPAPYKALSTSRVVWECSSKWVVNSI. This peptide was synthesized and used to produce specific polyclonal antibodies (antipeptides). This antipeptide was subsequently used to functionalize the QCM quartz crystal gold surface, creating a highly specific biosensor for the detection of Ko in honey samples. Further, the use of a peptide biomarker represents a novel application within the context of biosensor technology, marking an advancement over previously reported methods that lack specificity or require laboratory-based analyses.

3.4. Validation of Surface Immobilization

The first step, the successful deposition of the protein A layer, was confirmed by a decrease in the resonance frequency of the crystal, which ranged between 900 and 2500 Hz, indicating a mass increase in the three repeated experiments with 5 MHz crystals. Using a protocol akin to that used for protein A deposition, 200 μL of a stock solution of the antipeptide was applied. The successful deposition of the antipeptide was also confirmed by the decrease in the resonance frequency of the crystal, which ranged between 780 and 1500 Hz, indicating an increase in mass for the stable anchoring of the antipeptide.

3.5. Detection of Kodamaea ohmeri from the Biosensor

The peptide selected for this study was identified through a molecular biology study on various SHB-infested honey and hive product samples; details are reported in Section 3.1. This analysis validated the peptide identified as a representative biomarker of SHB infestation. The QCM instrument measures the frequency shift after peptide deposition on the antipeptide-functionalized quartz, indicating that a mass increase is attributable to the formation of the antipeptide–peptide complex. Approximately ten to hundreds of micrograms of peptide were detected, confirming the successful design of the strategy. Furthermore, the specific adhesion of the target substance to the substrate at the end of the rinsing procedure was established, thereby confirming the biosensor's ability to precisely identify Ko in honey samples. The control samples treated with PBS alone exhibited negligible frequency variations, confirming that the significant changes observed in the peptide-treated samples are due to the specific recognition of the biological entities of analytical interest.

The pivotal outcome of this methodology is depicted in Figure 3, which illustrates the frequency drop over time, a reduction attributed to the accumulation of mass on the biosensor surface due to the successful detection of the target peptide from Ko.

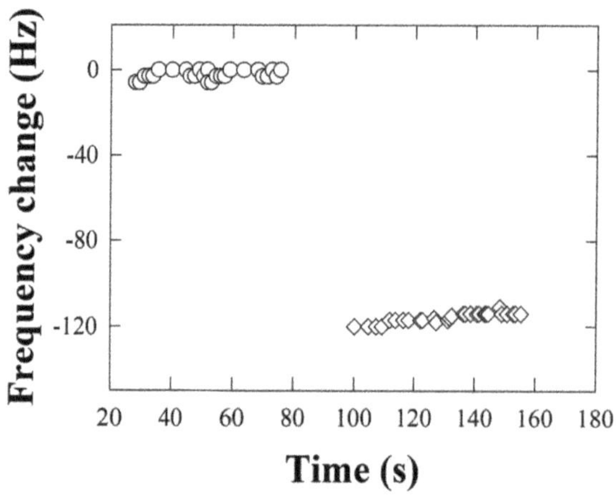

Figure 3. Changes in the frequency of oscillation of the quartz crystal correspond to the formation of the antipeptide–peptide complex. Open circles: frequency measurements of the functionalized quartz crystal before the addition of the peptide, taken as a zero reference to report the frequency shift. Open diamonds: measurements after peptide capture on the crystal quartz sensor.

Control experiments conducted under identical conditions, but in the absence of the target peptide using PBS alone, demonstrated negligible frequency shifts, as evidenced by the stability of the data points in the control set within Figure 3. The suggested method's effectiveness in identifying the presence of the Ko-associated peptide was demonstrated by the notable drop in frequency seen in the experimental set that was exposed to the peptide solution.

In essence, Figure 3 validates the utility of the developed microgravimetric biosensor technique for identifying the Ko peptide. The observed frequency reduction reflects the increase in mass deposition on the biosensor due to the captured peptide, thereby corroborating the effectiveness of the peptide-specific antibody in biosensor design.

Our biosensor utilizes QCM technology, which is effective at biomolecule detection due to its sensitivity to minute mass variations. This concept is employed to detect a specific peptide originating from the yeast Ko, serving as an early warning indicator for potential SHB infestation. The goal was to achieve excellent specificity for the detection of Ko. Since Ko is often found in SHB-infested hives, a sensor that detects its presence can be useful as a prevention system and should be seen as part of a broader toolkit for hive monitoring. While the present research does not focus on food safety evaluations, should future assessments determine that *K. ohmeri* in honey poses a food safety concern, the biosensor could be utilized for this distinct purpose. In this instance, additional regulations would be necessary for the proper implementation of the biosensor.

4. Conclusions

This research introduces an advanced biosensor for honeybee health monitoring, specifically designed to detect the Ko yeast as an indicator of SHB. It is highly precise, portable, and possesses significant potential for practical application in the field. Additional research is required to enhance sensitivity, elevate early detection thresholds, optimize robustness against varying environmental conditions, and minimize the costs of these devices for rapid dissemination. The integration of the current biosensor into standard practice could enable stakeholders, including beekeepers, honey producers, and local governments, to proactively combat SHB, thereby enhancing the resilience of pollination systems and ultimately safeguarding agricultural productivity and biodiversity.

The development of biosensors facilitated the acquisition of a method with high specificity through the meticulous selection of the target peptide, the immobilization strategy, and the application of QCM technology. The QCM-based biosensor is a viable alternative that requires additional research for optimization before implementation. Subsequent studies must focus on the enhancement of the sensor's sensitivity, thereby reinforcing its efficacy in early detection. The incorporation of this approach into beekeeping practices can facilitate proactive colony management and may mitigate the detrimental effects of delayed alerts regarding SHB infestation.

Author Contributions: Conceptualization, P.G. (Paola Ghisellini), P.G. (Patrizia Garbati), M.P., A.C., G.P., C.R., L.G., S.O., G.F. and R.E.; Methodology, P.G. (Paola Ghisellini), P.G. (Patrizia Garbati)., M.P., A.C., G.P., C.R., L.G., S.O., G.F. and R.E.; Validation, P.G. (Paola Ghisellini), P.G. (Patrizia Garbati)., M.P., A.C., G.P., C.R., L.G., S.O., G.F. and R.E.; Formal analysis, P.G. (Paola Ghisellini), P.G. (Patrizia Garbati)., M.P., A.C., G.P., C.R., L.G., S.O., G.F. and R.E.; Investigation, P.G. (Paola Ghisellini), P.G. (Patrizia Garbati)., M.P., A.C., G.P., C.R., L.G., S.O., G.F. and R.E.; Resources, G.F. and R.E.; Data curation, P.G. (Paola Ghisellini), P.G. (Patrizia Garbati)., M.P., A.C., G.P., C.R., L.G., G.F. and R.E.; Writing—original draft preparation, P.G. (Paola Ghisellini), P.G. (Patrizia Garbati)., M.P., A.C., G.P., C.R., L.G., S.O., G.F. and R.E.; Writing—review and editing, P.G. (Paola Ghisellini), P.G. (Patrizia Garbati)., M.P., A.C., G.P., C.R., L.G., S.O., G.F. and R.E.; Visualization, P.G. (Patrizia Garbati)., M.P., A.C., G.P., C.R., G.F. and R.E.; Supervision, G.F. and R.E.; Project administration, P.G. (Paola Ghisellini), L.G., G.F. and R.E.; Funding acquisition, P.G. (Paola Ghisellini), L.G., G.F. and R.E. All authors have read and agreed to the published version of the manuscript.

Funding: This publication is part of the project "New indicators and on-farm practices to improve honey bee health in the *Aethina tumida* Era in Europe" (BPRACTICES), which was funded under the ERA-NET framework of the European Union (Grant Agreement no. 69623).

Institutional Review Board Statement: The present work was performed under the guidance and request of the Ministero della Salute, General Direction of Animal Health and Veterinary Medicines (N. prot. DGSAF in Docspa/PEC). The animal study protocol was approved by the Institutional Review Board of the Ministry of Health, General Directorate of Animal Health and Veterinary Medicines, Office III-Animal Health and Operational Management of the National Center for the Fight and Emergency Against Animal Diseases and Unit Crisis Centre, Viale Giorgio Ribotta 5-00144 Rome, Italy (protocol code 0004352-21/02/2022-DGSAF-MDS-P and approval date 3 September 2022) for studies involving animals.

Data Availability Statement: The original contributions presented in this study are included in the article. Further inquiries can be directed to the corresponding author.

Conflicts of Interest: The authors declare that they have no known competing financial interests or personal relationships that could have appeared to influence the work reported in this paper.

Abbreviations

The following abbreviations are used in this manuscript:

SHB	Small Hive Beetle, *Aethina tumida*
Ko	*Kodamaea ohmeri*
QCM	Quartz Crystal Microbalance
COI	Cytochrome Oxidase I
ITS2	Internal Transcribed Spacer

References

1. Roth, M.A.; Wilson, J.M.; Gross, A.D. Biology and management of small hive beetles (Coleoptera: Nitidulidae): A pest of European honey bee (Hymenoptera: Apidae) colonies. *J. Integr. Pest Manag.* **2022**, *13*, 7. [CrossRef]
2. Mutinelli, F.; Montarsi, F.; Federico, G.; Granato, A.; Ponti, A.M.; Grandinetti, G.; Ferrè, N.; Franco, S.; Duquesne, V.; Rivière, M.-P.; et al. Detection of *Aethina tumida* Murray (Coleoptera: Nitidulidae.) in Italy: Outbreaks and early reaction measures. *J. Apic. Res.* **2014**, *53*, 569–575. [CrossRef]
3. Patel, V.; Pauli, N.; Biggs, E.; Barbour, L.; Boruff, B. Why bees are critical for achieving sustainable development. *Ambio* **2021**, *50*, 49–59. [CrossRef] [PubMed]
4. Torto, B.; Boucias, D.G.; Arbogast, R.T.; Tumlinson, J.H.; Teal, P.E.A. Multitrophic interaction facilitates parasite–host relationship between an invasive beetle and the honey bee. *Proc. Natl. Acad. Sci. USA* **2007**, *104*, 8374–8378. [CrossRef] [PubMed]
5. Liu, Y.; Beaurepaire, A.; Rogers, C.W.; Lopez, D.; Evans, J.D.; Straub, L.; Neumann, P.; Cook, S.C.; Huang, Q. Gene expression and functional analyses of odorant receptors in small hive beetles (*Aethina tumida*). *Int. J. Mol. Sci.* **2020**, *21*, 4582. [CrossRef] [PubMed]
6. Schäfer, M.O.; Cardaio, I.; Cilia, G.; Cornelissen, B.; Crailsheim, K.; Formato, G.; Lawrence, A.K.; Le Conte, Y.; Mutinelli, F.; Nanetti, A.; et al. How to slow the global spread of small hive beetles, *Aethina tumida*. *Biol. Invasions* **2019**, *21*, 1451–1459. [CrossRef]
7. Amos, B.A.; Leemon, D.L.; Hayes, R.A.; Cribb, B.W.; Furlong, M.J. Associations between the small hive beetle and the yeast *Kodamaea ohmeri* throughout the host life cycle. *J. Econ. Entomol.* **2018**, *111*, 1501–1508. [CrossRef] [PubMed]
8. Wilson, G.S.; Gifford, R. Biosensors for real-time in vivo measurements. *Biosens. Bioelectron.* **2005**, *20*, 2388–2403. [CrossRef] [PubMed]
9. Zhou, Q.; Zheng, C.; Zhu, L.; Wang, J. A review on rapid detection of modified quartz crystal microbalance sensors for food: Contamination, flavour and adulteration. *TrAC Trends Anal. Chem.* **2022**, *157*, 116805. [CrossRef]
10. Li, D.; Waite, D.W.; Fan, Q.H.; George, S.; Semeraro, L.; Blacket, M.J. Molecular detection of small hive beetle *Aethina tumida* Murray (Coleoptera: Nitidulidae): DNA barcoding and development of a real-time PCR assay. *Sci. Rep.* **2018**, *8*, 9623. [CrossRef] [PubMed]
11. Santino, I.; Bono, S.; Borruso, L.; Bove, M.; Cialdi, E.; Martinelli, D.; Alari, A. *Kodamaea ohmeri* isolate from two immunocompromised patients: First report in Italy. *Mycoses* **2013**, *56*, 179–181. [PubMed]
12. Ward, L.; Brown, M.; Neumann, P.; Wilkins, S.; Pettis, J.; Boonham, N. A DNA method for screening hive debris for the presence of small hive beetle (Aethina tumida). *Apidologie* **2007**, *38*, 272–280. [CrossRef]
13. Ghisellini, P.; Caiazzo, M.; Alessandrini, A.; Eggenhöffner, R.; Vassalli, M.; Facci, P. Direct electrical control of IgG conformation and functional activity at surfaces. *Sci. Rep.* **2016**, *6*, 37779. [CrossRef] [PubMed]
14. Pizzoni, D.; Mascini, M.; Lanzone, V.; Del Carlo, M.; Di Natale, C.; Compagnone, D. Selection of peptide ligands for piezoelectric peptide based gas sensors arrays using a virtual screening approach. *Biosens. Bioelectron.* **2014**, *52*, 247–254. [CrossRef] [PubMed]

insects

MDPI

Article

Chronic Larval Exposure to Lambda-Cyhalothrin Alters Gene Expression in Both Larval and Adult Honey Bees (*Apis mellifera*)

Bala Murali Krishna Vasamsetti, Kyongmi Chon *, Juyeong Kim, Minju Choi, Bo-Seon Kim, Chang-Young Yoon, Sojeong Hwang and Kyeong-Hun Park

Toxicity and Risk Assessment Division, Department of Agro-Food Safety and Crop Protection, National Institute of Agricultural Sciences, Rural Development Administration, Iseo-myeon, Wanju-gun 55365, Jeollabuk-do, Republic of Korea; vbmk84@gmail.com (B.M.K.V.); kjy.sara@gmail.com (J.K.); chlalswn1102@gmail.com (M.C.); kbs9249@naver.com (B.-S.K.); evermoo2600@korea.kr (C.-Y.Y.); hsj102@korea.kr (S.H.); blueour@korea.kr (K.-H.P.)

* Correspondence: kmchon6939@korea.kr

Simple Summary: Honey bees are crucial for pollinating crops and maintaining balanced ecosystems, but their populations are declining around the world. One main cause is exposure to pesticides used in agriculture. In this study, we explored the impacts of lambda-cyhalothrin (LCY), a common insecticide used to control pests in crops. We exposed honey bee larvae to a very low dose of LCY and investigated its effects on their gene expression. The pesticide weakened their ability to defend against stress and disrupted important biological processes in the larvae. Exposure at the larval stage also showed toxic effects in adult bees, suggesting that even low levels of LCY can cause lasting harm. We found that it altered the activity of genes involved in development, energy metabolism, and brain function at both developmental stages. Our findings highlight the importance of implementing safer farming practices to protect pollinators.

Abstract: Lambda-cyhalothrin (LCY), a widely used pyrethroid insecticide, is toxic to bees—vital pollinators experiencing global declines; however, its molecular effects during early development remain poorly understood. We investigated the molecular mechanisms underlying chronic sublethal exposure to LCY in the larval and adult stages. Larvae were exposed to LCY (0.004 μg active ingredient/larva), with four groups examined: solvent-treated larvae group (SLG), solvent-treated adult group (SAG), LCY-treated larvae group (LLG), and LCY-treated adult group (LAG). We identified 1128 and 168 significantly altered genes in LLG vs. SLG and LAG vs. SAG, respectively, with 125 larval- and 25 adult-specific DEGs, indicating stage-dependent toxicity. LCY dysregulated processes such as cuticle formation, sulfur metabolism, oxidoreductase activity, and neuropeptide signaling in larvae, while adults exhibited altered redox balance, peptide receptor signaling, and monoamine transport. Neuroactive signaling disruptions were observed in both stages, with additional effects on motor function, amino acid metabolism, and glycolysis in larvae; whereas adults exhibited altered lipid biosynthesis and energy metabolism. Downregulated genes involved in chitin metabolism and antioxidant defenses in larvae suggested compromised exoskeletal integrity and increased vulnerability. Overall, our findings highlight the long-term molecular consequences of early-life exposure and emphasize the need for safer pesticide practices to protect pollinator health.

Keywords: *Apis mellifera*; lambda-cyhalothrin; transcriptomic analysis; sublethal pesticide toxicity; Gene Ontology; KEGG pathway analysis

1. Introduction

Honey bees play a critical role in maintaining the ecological balance through the pollination of various plants, making them essential for ecosystem stability [1]. Beyond their ecological significance, honey bees contribute valuable products, such as honey, royal jelly, and propolis, which have various nutritional, medicinal, and cosmetic applications. The decline in honey bee populations has emerged as a global concern, prompting significant research efforts to identify the reasons for this reduction [2,3]. Among the various factors contributing to this decline, pesticide exposure stands out as a major threat to honey bee health and survival [4–6].

During foraging, honey bees collect nectar and pollen that may be contaminated with pesticides and bring them back to the hive. Inside the hive, this contaminated nectar and pollen are transformed into honey and bee bread, which nurse bees feed to the larvae along with royal jelly, thereby exposing them to pesticide residues [5,7–10]. Such exposure can disrupt critical behaviors such as foraging, navigation, and learning, and can weaken the immune systems of bees, making them more susceptible to additional stressors, such as pathogens and habitat loss [4,10]. Notably, pesticide exposure has been shown to inhibit acetylcholinesterase activity, alter antioxidant enzyme levels such as those of catalase and superoxide dismutase, and disrupt the expression of genes related to neural signaling and immune regulation [10,11]. These physiological disruptions compromise the bees' ability to cope with additional stressors, potentially contributing to colony collapse disorder and declines in honey bee populations [3,4,10]. Both larvae and adult bees are vulnerable to pesticide exposure, and studies have reported significant effects on survival, cognitive ability, and overall colony dynamics [12–15]. While neonicotinoids have been widely scrutinized and restricted owing to their off-target toxicity, pyrethroids are increasingly used as alternative pesticides, raising new concerns about their impact on honey bee health [11,16–19].

Lambda-cyhalothrin (LCY), a pyrethroid insecticide, has become widely used. Commonly applied to agricultural crops, including almonds, apples, and cherries, it protects against insect pests such as aphids, caterpillars, and beetles [20]. LCY acts on the nervous system of insects by altering voltage-gated sodium, calcium, and chloride channels [21]. Beyond its intended targets, LCY poses toxic risks to humans and non-target animals. In humans, in vitro exposure to LCY has been associated with chromosomal aberrations and DNA damage in lymphocytes, indicating genotoxic potential [22]. LCY also disrupts antioxidant enzyme activities and induces oxidative stress in erythrocytes [22,23]. In human astrocytes, it disrupts calcium homeostasis and triggers apoptosis [24]. Animal studies further revealed hepatotoxicity in rats, with toxicity increasing in a dose- and time-dependent manner but mitigated by ascorbic acid [25–27]. The underlying mechanism of LCY toxicity involves the disruption of ionic conduction across sodium, calcium, and chloride channels, leading to oxidative tissue damage [27,28], highlighting its potential for neurotoxicity, oxidative damage, and genomic changes across a range of non-target organisms.

Environmentally, LCY residues have been detected in pollen at concentrations ranging from 3.2 to 1750 μg/kg, and at 314.3 μg/kg in bee bread [10,29,30]. LCY has been shown to cause severe toxicological effects in non-target pollinators like honey bees, even at sublethal concentrations. Acute exposure significantly reduces foraging behavior, impairs learning and memory, and affects homing ability, indicating neurotoxic impacts [31,32]. Furthermore, LCY exposure during the larval stage increases mortality, causes deformities, and disrupts key developmental enzymes [11]. Laboratory assessments confirmed that both contact and oral exposure led to high mortality rates, with LD_{50} values highlighting substantial toxicity to *Apis cerana indica* [14]. Field assessments provided additional insights; semi-field and field studies found that LCY exposure (contact) reduces colony foraging activity, adversely affects brood development, and depletes stored nectar and pollen reserves [15]. Additionally, research has shown that LCY alters the gut microbiota, immune responses, and detoxification pathways, compounding concerns about its physiological impacts [6,11,33]. Furthermore, LCY disrupts foraging activities, shortens lifespan, impairs memory, and damages critical bee organs, such as the brain, hypopharyngeal glands, and midgut [32,34,35].

The Pesticide Properties Database lists LCY as an insecticide with notable toxicity to honey bees, with an acute contact LD_{50} of 0.038 μg/bee and an acute oral LD_{50} of 0.91 μg/bee [36]. Our previous research established the acute and chronic LD_{50} values of LCY in honey bee larvae as 0.058 μg a.i./larva and 0.040 μg a.i./larva, respectively [11]. These lower values compared to the adult oral LD_{50} indicate that larvae are more sensitive to LCY exposure. Furthermore, sublethal larval exposure to LCY causes wing and antenna deformities in emerged adult bees [11], pointing to lasting developmental toxicity. This highlights the need for further investigation into the molecular mechanisms underlying LCY's impact on honey bee development.

Toxicity assessments in honey bees often focus on mortality, phenotypic changes (antenna and wing deformities), and behavioral alterations [11,12,37]. Given the chemical complexity of pesticides, the phenotypic changes observed in honey bees following exposure are likely driven by intricate molecular disruptions that traditional techniques, such as ELISA and qRT-PCR, fail to fully elucidate. Considering these limitations, next-generation whole-transcriptome sequencing offers a powerful approach to uncover complex molecular changes [38–41]. To broaden our understanding of the molecular pathways involved in LCY-induced toxicity, we used this technology to analyze global transcriptomic changes in honey bees following chronic larval LCY exposure. Differentially expressed genes (DEGs) were identified and validated using qRT-PCR, followed by functional enrichment analyses to uncover the mechanistic basis of LCY-induced toxicity.

2. Materials and Methods

2.1. Larval Exposure to LCY

2.1.1. In Vitro Rearing of First-Instar Honey Bee Larvae

First-instar honey bee (*Apis mellifera*) larvae used in the experiments were obtained from an experimental apiary at the National Institute of Agricultural Sciences, Rural Development Administration, Republic of Korea (35.591° N, 126.278° E). The colonies were maintained in a controlled environment for four weeks prior to the study, ensuring that no chemical substances were applied to the hives during this period. Procedures for larval collection, grafting, feeding, and experimental conditions followed established protocols [11,12]. Briefly, to obtain age-matched larvae, the queen was confined to an empty comb for 24 h to induce egg laying. After 72 h, combs were transferred to the laboratory, and first-instar larvae were grafted into sterile 48-well plates (#32024, SPL, Pocheon-si, Republic of Korea) and incubated at 35 °C, 95 ± 5% relative humidity. The larval diet, consisting of royal jelly, glucose, fructose, yeast extract, and distilled water, was freshly prepared, stored at 4 °C, and brought to 35 °C prior to feeding. For detailed protocols, refer to the Supplementary File.

2.1.2. Chronic Sublethal Larval LCY Exposure, and Collection of Larval and Adult Honey Bee Samples

LCY, with a purity of 95%, sourced from Sigma (St. Louis, MO, USA), was initially prepared in acetone and mixed with the diet to obtain the desired concentration. Honey bee larvae were subjected to LCY in accordance with the protocols of previous studies [11,12] (see Supplementary Files for more details). The dosage of LCY chosen for larval exposure was 0.004 μg a.i./larva, representing one-tenth of the chronic LD_{50} value as previously established in our laboratory [11]. From days 3 to 6 post-grafting, each larva was exposed to 0.004 μg LCY/larva. The acetone concentration was maintained at 0.5% in the LCY exposure group. The LCY exposure was performed in triplicate, with each replicate containing 36 larvae. In the control group, the larvae were fed a diet containing 0.5% acetone. Dead larvae were removed daily. Larvae were collected on day 8 after grafting, and newly emerged adult bees were collected from day 19. For each experimental condition, three replicate samples were collected, with each sample consisting of three individual larvae or adults. The samples were rapidly frozen in liquid nitrogen and stored at −80 °C. Separate sets of samples were collected for transcriptomic analysis (n = 3) and qRT-PCR validation (n = 3). In total, four experimental groups were established based on treatment type and developmental stage: the solvent-treated larvae group (SLG), the solvent-treated adult group (SAG), the LCY-treated larvae group (LLG), and the LCY-treated adult group (LAG).

2.2. *Transcriptome Analysis*

2.2.1. cDNA Library Construction

RNA isolation, RNA quantification, RNA integrity assessment, and cDNA library construction followed previously established methodologies [42], with comprehensive procedures provided in the Supplementary File. Briefly, total RNA was extracted using the RNeasy® Mini Kit (#74106; QIAGEN, Hilden, Germany). RNA quality was confirmed using an RNA Screentape (#5067-5576; Agilent, Santa Clara, CA, USA), and samples with an RNA integrity number (RIN) >7 were used for cDNA library preparation.

The indexed cDNA libraries were subjected to sequencing on the NovaSeq platform by Illumina, Inc. (San Diego, CA, USA) using a paired-end setup (2 × 100 bp). To improve data quality, adapter sequences were removed, and low-quality bases were trimmed employing Trimmomatic v0.38 [43]. Subsequently, the cleaned reads were mapped to the *A. mellifera* genome (Amel_HAv3.1) utilizing HISAT v2.1.0, which was developed upon the frameworks of HISAT and Bowtie2 [44]. After alignment, reads were processed with SAMtools v1.9 [45]. Transcript assembly and quantification, including raw counts, FPKM, and TPM, were performed with StringTie v2.1.3b [46].

2.2.2. DEG Filtering

DESeq2 v1.24.0 [47] was used to analyze differential gene expression, with raw count data serving as input. During quality control, genes with nonzero counts across all replicates of at least one group were retained. Principal component analysis and multidimensional scaling plots were generated to assess sample clustering and expression pattern consistency. To account for differences in library size, relative log expression normalization was applied. Differential expression analysis was performed using DESeq2's negative binomial Wald test, yielding fold changes (FCs) and p-values [47]. To minimize false positives, p-values were adjusted using the Benjamini–Hochberg method for false discovery rate control. Genes were considered significantly differentially expressed if they met the criteria of $|FC| \geq 1.5$ and raw p-value < 0.05. Hierarchical clustering of significant genes was carried out on log-transformed data, employing the Euclidean distance as the metric and complete linkage as the clustering method.

2.2.3. DEG Enrichment Analysis

Functional annotation and gene enrichment analyses of significant genes were performed using gProfiler [48] by referencing the Gene Ontology (GO) and Kyoto Encyclopedia of Genes and Genomes (KEGG) pathway databases. Statistical significance was assessed using a one-sided hypergeometric test, and *p*-values were adjusted using the Benjamini–Hochberg correction to control for false discovery rate.

2.3. DEG Validation

RNA extraction, cDNA synthesis, and qRT-PCR were conducted as previously described [42]. Briefly, RNA was extracted from 50 mg larval or adult tissue using TRIzol Plus (#PG1117; Progen Life Sciences, New York, NY, USA) and the Direct-zol RNA Miniprep Kit (#R2052; Zymo research, Irvine, CA, USA), following the manufacturer's protocols. RNA quality was verified using a Nanodrop 2000 (Thermo Fisher Scientific, Waltham, MA, USA), and 1500 ng of RNA was reverse-transcribed into cDNA using a kit (#K2205; Bioneer, Daejon, Republic of Korea). qRT-PCR analyses were conducted using the K6251 kit from Bioneer (Daejeon, Republic of Korea) on a CFX96 Dx system (Bio-Rad, Hercules, CA, USA). The fold differences in expression were determined via the $2^{-\Delta\Delta CT}$ method [49], utilizing the average CT values from three housekeeping genes (*RPL13a* (*LOC108003218*), *GAPDH* (*LOC108001468*), and *RPS5* (*LOC107998344*)) for improved normalization. For detailed protocols and oligonucleotide primer sequences (Table S1), refer to the Supplementary File. Statistical analysis of qRT-PCR data was conducted using the Mann–Whitney U test in GraphPad Prism software (version 6, GraphPad Software, Boston, MA, USA).

A schematic overview of the workflow, including larval grafting, LCY exposure, staged sampling, and transcriptome analysis, is provided in Figure S1.

3. Results

3.1. DEGs in Honey Bee Larvae and Adults Exposed to LCY

To identify DEGs in response to chronic sub-lethal LCY exposure, paired-end transcriptome sequencing was conducted using the Illumina platform across the four experimental groups: SLG, SAG, LLG, and LAG. Overall, 56.21 million raw reads were generated. After quality control, 97.47% of the reads were retained as clean reads, yielding an average of 54.79 million clean reads across all samples (Table 1). The guanine and cytosine (GC) content of the clean reads was approximately 38.43%, and the quality scores were notably high, with Q20 at 98.65% and Q30 at 97.70%. The high-quality reads were mapped to the *A. mellifera* reference genome Amel_HAv3.1 (NCBI_20180910), achieving an average alignment rate exceeding 97.16% across all sample groups. At an FC $\geq$ 1.5 and a raw *p*-value < 0.05, the comparisons of SAG vs. SLG, LAG vs. LLG, LLG vs. SLG, and LAG vs. SAG yielded 5687, 5789, 1128, and 168 DEGs, respectively (Figure 1). The list of DEGs across all comparisons is included in Supplementary File S1. Heat maps were constructed using two-way hierarchical clustering based on the Z-scores of log2-transformed normalized values, allowing visualization of gene expression similarities and differences across samples (Figure S2). Volcano plots were generated for each comparison (Figure S3), identifying significantly upregulated and downregulated genes, with the most pronounced changes appearing in the upper-right (upregulated) and upper-left (downregulated) regions.

Table 1. Summary of transcriptome data, including read counts, sequence quality, and mapping rates.

Samples	Total Reads	Clean Reads	GC (%)	Q20 (%)	Q30 (%)	Mapped Reads	Unmapped Rate (%)	Mapped Rate (%)
LAG-1	58,406,928	56,777,182	38.40	98.48	97.65	55,333,933	2.54	97.46
LAG-2	53,953,234	52,345,476	38.85	98.46	97.62	50,921,510	2.67	97.28
LAG-3	54,812,774	53,237,224	38.06	98.49	97.64	51,798,939	2.70	97.30
LLG-1	56,396,336	55,070,232	39.61	98.79	97.68	53,431,764	2.98	97.02
LLG-2	50,705,676	49,305,624	39.42	98.55	97.73	47,920,635	2.81	97.19
LLG-3	60,087,126	58,930,328	39.39	98.91	97.95	57,267,666	2.82	97.18
SAG-1	62,326,854	60,905,516	37.71	98.75	97.73	59,163,527	2.86	97.14
SAG-2	57,545,836	56,071,312	37.18	98.72	97.48	54,287,287	3.18	96.82
SAG-3	55,289,460	54,121,612	37.86	98.83	97.78	52,478,090	3.04	96.96
SLG-1	62,347,274	60,872,770	37.62	98.77	97.70	59,142,506	2.84	97.16
SLG-2	50,206,390	48,908,502	38.75	98.53	97.73	47,621,910	2.63	97.37
SLG-3	52,455,156	50,888,552	38.27	98.50	97.67	49,363,367	3.00	97.00

SLG, solvent-treated larvae group; SAG, solvent-treated adult group; LLG, LCY-treated larvae group; LAG, LCY-treated adult group; GC, guanine and cytosine content; Q20, proportion of bases with a quality score of 20 or higher; Q30, proportion of bases with a quality score of 30 or higher.

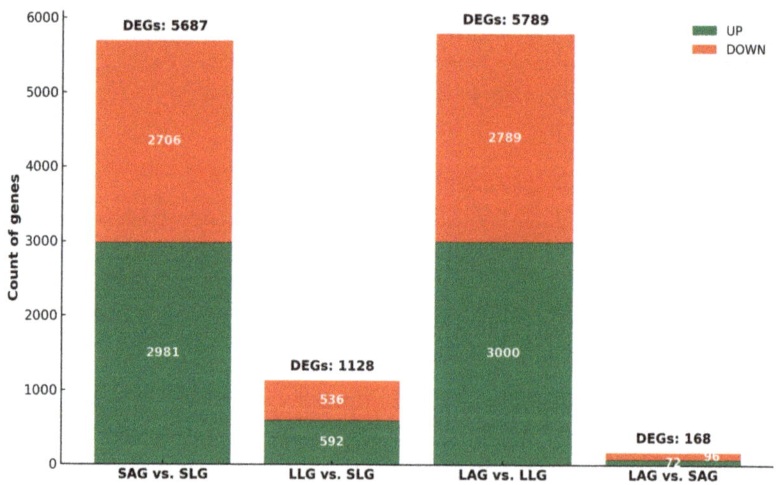

Figure 1. Differential gene expression across experimental conditions. The bar chart illustrates the number of differentially expressed genes (DEGs) (|FC| $\geq$ 1.5, $p < 0.05$), classified as upregulated (green) or downregulated (red), across four comparisons: SAG vs. SLG, LLG vs. SLG, LAG vs. LLG, and LAG vs. SAG. SLG (solvent-treated larvae group), SAG (solvent-treated adult group), LLG (LCY-treated larvae group), and LAG (LCY-treated adult group).

3.2. Distinct and Overlapping DEGs in LCY-Exposed Honey Bee Larvae and Adult Groups

A Venn diagram was generated to examine the overlap and distinctiveness of the DEGs among the experimental groups (Figure 2). The analysis identified 4896 DEGs shared between the SAG vs. SLG and LAG vs. LLG comparisons, emphasizing their essential roles in the larval-to-adult transition and normal honey bee development. In addition, 791 DEGs were specific to the SAG vs. SLG comparison, whereas 893 DEGs were exclusive to the LAG vs. LLG comparison.

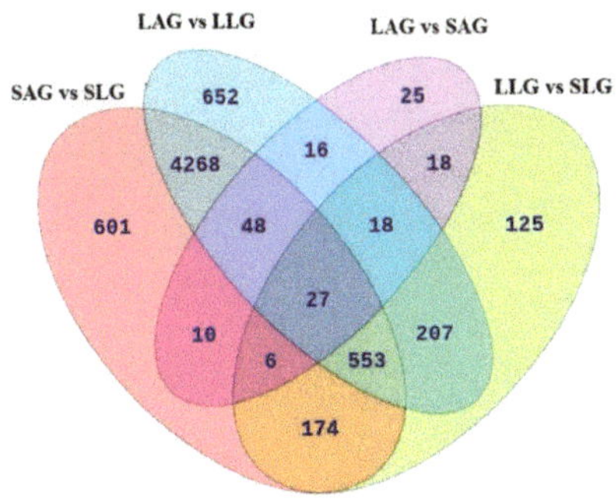

Figure 2. Venn diagram depicting shared and unique DEGs among experimental groups. The Venn diagram displays the unique and shared DEGs among the following comparisons, offering insights into gene expression alterations across different treatments: SAG vs. SLG, LAG vs. LLG, LLG vs. SLG, and LAG vs. SAG. SLG (solvent-treated larvae group), SAG (solvent-treated adult group), LLG (LCY-treated larvae group), and LAG (LCY-treated adult group).

Following LCY exposure, 125 distinct DEGs were identified in larvae, including genes linked to stress response (e.g., *hsp70*, *cyp6as14*) and neurodevelopment (e.g., *map2*, *Sog*). In adults, 25 unique DEGs were identified, such as glucose dehydrogenase [*FAD*, quinone], and deleted in autism protein 1 homolog, which are implicated in energy metabolism and neural function. A complete list of shared and condition-specific DEGs is provided in Supplementary File S2.

3.3. GO Enrichment of Significant DEGs

GO enrichment analysis of the significant DEGs was conducted to categorize them into three groups: biological processes (BPs), cellular components (CCs), and molecular functions (MFs).

3.3.1. LLG vs. SLG

The top enriched MFs included structural constituents of the cuticle, sulfate transmembrane transporter activity, secondary active sulfate transmembrane transporter activity, monooxygenase activity, sulfur compound transmembrane transporter activity, oxidoreductase activity, FMN binding, neuromedin U receptor activity, oxidoreductase activity acting on paired donors, and neuropeptide receptor activity (Figure 3A). These enriched functions point to significant alterations in cuticle formation, transport systems, redox reactions, and signaling pathways, suggesting a potential impact of LCY on larval development and physiology. Notably, genes related to cuticle structure (e.g., *cpr* family genes) and sulfate transporters (e.g., *LOC413816*, *LOC410507*), as well as multiple detoxification-related genes such as *Cyp314a1* and *nos*, were significantly affected. The genes related to these terms are listed in Table S2.

3.3.2. LAG vs. SAG

The top enriched MFs included tetrapyrrole binding, G protein-coupled peptide receptor activity, peptide receptor activity, heme binding, flavin adenine dinucleotide binding, oxidoreductase activity, oxidoreductase activity acting on the CH–OH group of donors, monoamine transmembrane transporter activity, and methylenetetrahydrofolate reductase (NAD(P)H) activity (Figure 3B). These functions indicate disruptions in binding activities, redox reactions, and signaling pathways, which may have lasting developmental consequences extending into adulthood, potentially influencing neural function and behavior. Notably, genes involved in redox reactions and cofactor binding (e.g., *LOC408452*,

LOC551179, *LOC551044*) and neuropeptide signaling (e.g., *akhr*, *LOC411672*) were differentially expressed. Additionally, *LOC100578936*, associated with folate metabolism, may reflect altered one-carbon metabolism under LCY exposure. These results suggest that LCY exposure changes neurometabolic gene expression, potentially leading to lasting behavioral and neurological issues in adult bees. The list of genes for these terms is included in Table S3.

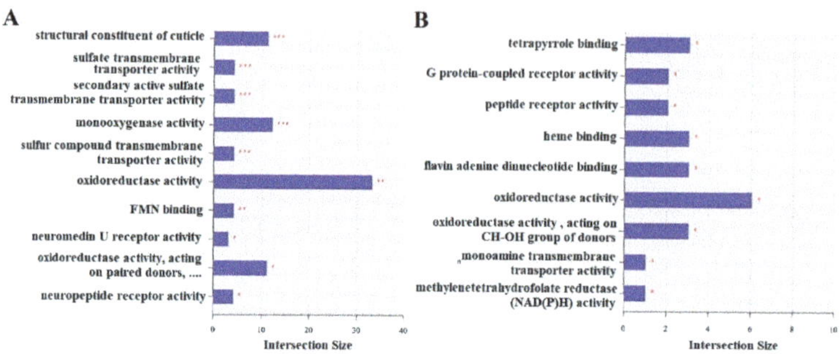

Figure 3. Top enriched GO terms for molecular function in DEGs between treatment groups. (**A**) Top 10 enriched MFs in LLG vs. SLG. (**B**) Top nine enriched BPs in LAG vs. SAG. SLG: solvent-treated larvae group; SAG: solvent-treated adult group; LLG: LCY-treated larvae group; LAG: LCY-treated adult group. Statistical significance: * $p < 0.05$, ** $p < 0.01$, *** $p < 0.001$.

3.3.3. LAG vs. LLG and SAG vs. SLG

A significant number of DEGs overlapped in the SAG vs. SLG and LAG vs. LLG comparisons (Figure 2), which led to shared GO categories.

In the CCs category, eight of the top ten terms were shared across both comparisons (Figure S4), highlighting the key cellular structures essential for mitochondrial function and energy homeostasis. While the respirasome and proton-transporting two-sector ATPase complexes were exclusive to the SAG vs. SLG comparison, the cytoplasmic and organelle membranes were specific to the LAG vs. LLG comparison. These findings suggest that LCY exposure may differentially affect mitochondrial integrity and cellular compartmentalization during honey bee development.

In the MFs category, eight of the top ten terms overlapped between the comparisons (Figure S5), highlighting the importance of membrane transport and structural integrity in honey bees. While proton transmembrane transporter activity and monoatomic ion transmembrane transporter activity were unique to the SAG vs. SLG comparison, electron transfer activity and oxidoreductase activity were specific to the LAG vs. LLG comparison. This indicates that LCY exposure may selectively alter membrane transport processes and redox homeostasis, potentially disrupting cellular functions during development.

In the BPs category, eight of the top ten terms were common across both comparisons (Figure S6), emphasizing the pathways that are critical for cellular energy balance, essential functions, and honey bee growth and development. Notably, cellular respiration and ATP synthesis coupled electron transport were specific to the SAG vs. SLG comparison, whereas purine-containing compound metabolic process and nucleotide metabolic process were unique to the LAG vs. LLG comparison. This suggests that LCY exposure disrupts cellular energy production and genetic material metabolism, potentially affecting honey bee development from larvae to adults. The lists of genes associated with these CC, MF, and BP terms are shown in Tables S4–S9.

3.4. KEGG Enrichment Analysis

The enriched KEGG pathways were grouped into six major categories, and a full list of identified biological pathways is provided in Figure S7.

3.4.1. LLG vs. SLG

The top 10 enriched pathways in the LLG included neuroactive ligand–receptor interaction; motor proteins; ascorbate and aldarate metabolism; tyrosine metabolism; tryptophan metabolism; pentose and glucuronate interconversions; glycolysis/gluconeogenesis; phenylalanine metabolism; valine, leucine, and isoleucine degradation; and fatty acid degradation (Figure 4A). These results suggest that exposure to LCY causes significant disruptions in nervous system function, energy metabolism, oxidative stress response, and overall metabolic health, potentially affecting the normal growth and development of larvae. Specifically, LCY exposure appears to dysregulate genes involved in nervous system function (e.g., *nAChRa9*, *dop1*, *sifr*), energy metabolism (e.g., *tpi*, *LOC408559*), oxidative stress response (e.g., *LOC408650*, *LOC411140*), and overall metabolic health (e.g., *LOC551465*, *LOC551837*), potentially impacting larval development. A list of the genes for each pathway is presented in Table S10.

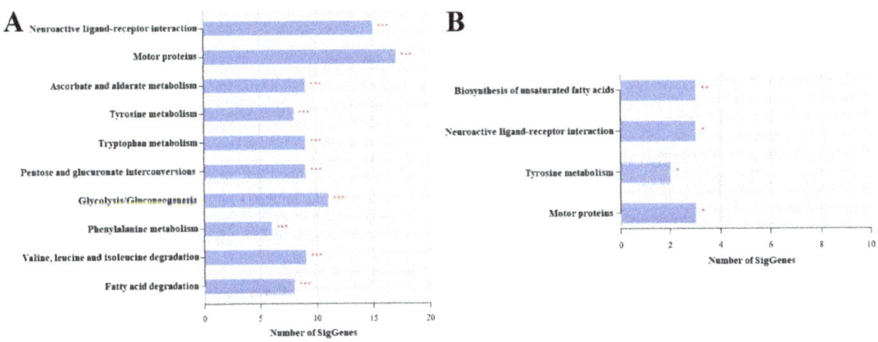

Figure 4. Top 10 enriched KEGG pathways across different transcriptome comparisons. The panels represent the most enriched KEGG pathways identified in the following comparisons: (**A**) LLG vs. SLG and (**B**) LAG vs. SAG. SLG: solvent-treated larvae group; SAG: solvent-treated adult group; LLG: LCY-treated larvae group; LAG: LCY-treated adult group. Statistical significance is indicated as follows: * $p < 0.05$, ** $p < 0.01$, *** $p < 0.001$.

3.4.2. LAG vs. SAG

The top enriched pathways in the LAG included the biosynthesis of unsaturated fatty acids, neuroactive ligand–receptor interaction, tyrosine metabolism, and motor proteins (Figure 4B). These enriched pathways suggest that LCY exposure may disrupt lipid metabolism, neural signaling, and muscle function, potentially impairing essential behaviors and colony health in adult honey bees. Specifically, genes such as *LOC727166* and *LOC724226*, involved in fatty acid synthesis, may influence energy storage and flight performance, whereas *akhr* and *nAChRa9*, linked to neuropeptide signaling and sensory function, and *LOC551465*, associated with neurotransmitter production, indicate neural impacts. Additionally, motor-related genes such as *LOC100578129* and *LOC551109* affect muscle function. Changes in the expression levels of these genes suggest that LCY exposure could impair individual performance and, cumulatively, affect colony-level stability and productivity. A list of genes that belong to each enriched pathway is presented in Table S11.

3.4.3. LAG vs. LLG and SAG vs. SLG

KEGG pathway analysis revealed key developmental pathways in both control and LCY-treated groups. In the controls (SAG vs. SLG), 10 pathways were enriched, reflecting metabolic and regulatory processes (Figure S8A). Upon LCY exposure (LAG vs. LLG), eight of these pathways remained enriched, whereas peroxisome function and nucleocytoplasmic transport were uniquely affected (Figure S8B), suggesting cellular stress and impaired molecular transport. In addition, neuroactive ligand–receptor interaction and pyruvate metabolism, which were enriched in the controls, were absent in the LCY-treated group, indicating disruptions in neurotransmission and energy balance. These results suggest that LCY exposure alters metabolic and signaling pathways beyond normal developmental changes, with potential long-term effects on physiology and behavior. The list of genes of each enriched pathway is included in Tables S12 and S13.

3.5. Transcriptome Data Validation

Three upregulated and three downregulated genes from each comparison were randomly selected and analyzed via qRT-PCR to validate the reliability of the transcriptome data. All genes tested using qRT-PCR exhibited similar trends to those observed in the transcriptome analysis data (Figure 5). Selected genes and their expression statistics are provided in Table S14. The direction and trend of gene expression were found to be similar in both qRT-PCR and transcriptome analysis. A scatter plot showed a strong correlation between transcriptome and qRT-PCR FCs (R^2 = 0.658) (Figure S9), confirmed by Pearson (r = 0.811, $p < 0.0001$) and Spearman (ρ = 0.951, $p < 0.0001$) analyses.

LAG vs LLG	Transcriptome	qRT PCR
cutP	101.69	33.94
d-galT	172.88	60.42
aGLU2	297.17	81.75
hex70c	-10,581.34	-15,148.76
lig1-X2	-1,639.88	-1,273.22
obp13	-511.49	-2,526.88

SAG vs SLG	Transcriptome	qRT PCR
tpnc3b	6,161.15	29,530.82
melt	3,772.67	17,718.17
LOC725364	7,017.85	55,761.43
hex110	-4,092.66	-560.85
fib4	-1,187.30	-4,956.91
vhdl	-2,338.50	-1,506.35

LLG vs SLG	Transcriptome	qRT PCR
pox	10.86	2.73
zncpt	9.4	1.18
cutP14	7.95	1.27
tert	-5.74	-1.65
orc5	-5.33	-1.71
yg	-5.18	-2.62

LAG vs SAG	Transcriptome	qRT PCR
gdh-fq-x2	2.79	1.07
ttc17ip	2.00	1.61
cyb561d2	2.79	1.54
elovl1	-5.82	-3.79
kif9	-3.47	-53.41
dnajc18l	-1.98	-3.00

Figure 5. Comparison of gene expression data between transcriptome analysis and qRT-PCR across honey bee sample comparisons. The graphs illustrate the fold changes (FCs) in expression levels of selected genes, with qRT-PCR results displayed alongside corresponding transcriptome data. SLG (solvent-treated larvae group), SAG (solvent-treated adult group), LLG (LCY-treated larvae group), and LAG (LCY-treated adult group). Gene abbreviations: *cutp* (Cuticular protein), *d-galt* (D-galactonate transporter), *aglu2* (Alpha-glucosidase 2), *hex70c* (Hexamerin 70c), *lig1-x2* (DNA ligase 1, variant X2), *obp13* (Odorant-binding protein 13), *pox* (Peroxidase), *zncpt* (Zinc carboxypeptidase), *cutp14* (Cuticular protein 14), *tert* (Telomerase reverse transcriptase), *orc5* (Origin recognition complex

subunit 5), *yg* (Yellow-g), *gdh-fq-x2* (Glucose dehydrogenase [FAD, quinone], variant X2), *melt* (Melittin), *LOC725364* (Flightin), *hex110* (Hexamerin 110), *fib4* (Silk fibronin 4), *vdhl* (Larva-specific very high-density lipoprotein), *ttc17ip* (TTC17-interacting protein), *cyb561d2* (Cytochrome b561 domain-containing protein 2), *elovl1* (Elongation of very long-chain fatty acids protein 1), *kif9* (Kinesin-like protein 9), and *dnajc18l* (DNAJ homolog subfamily C member 18-like). Blue indicates downregulation, and red indicates upregulation, of gene expression levels in LCY-treated samples relative to solvent-treated controls.

4. Discussion

The adverse effects of LCY on honey bees are well-documented, including increased mortality, morphological deformities in larvae, and behavioral disruptions in adults [11,14,15,31,32,50]. We employed next-generation transcriptome analysis to investigate LCY's molecular impact on bees, revealing that it interferes with crucial signaling pathways in both larvae and adults. In larvae, LCY disrupted carbohydrate and fatty acid metabolism and pathways related to structural integrity, thereby compromising energy homeostasis and normal growth. In adults, persistent neurotoxicity was evident, with enrichment in pathways related to neuroactive ligand–receptor interactions, tyrosine metabolism, and synaptic transmission impairments. Despite these differences, both groups exhibited systemic disruptions in oxidative stress and detoxification pathways, highlighting the pervasive impact of LCY toxicity. Collectively, these molecular alterations reinforce the link between LCY exposure and previously reported developmental and behavioral impairments in honey bees.

4.1. Metabolic and Neurodevelopmental Impacts of LCY Exposure in Honey Bee Larvae

Chronic exposure to LCY may disrupt crucial metabolic, neurological, oxidative, and developmental processes in larvae by dysregulating genes that are vital for energy metabolism, neurotransmitter biosynthesis, structural integrity, and detoxification. These molecular disruptions align with previously observed phenotypic and biochemical impairments, contributing to long-term developmental defects, diminished adult fitness, and overall colony decline [11,15,32,50].

Metabolic dysfunction appears to be a notable consequence of LCY exposure, as demonstrated by widespread transcriptional changes in key mitochondrial enzymes, detoxification systems, and energy metabolism pathways [51–54]. KEGG enrichment of glycolysis/gluconeogenesis, pentose and glucuronate interconversions, and fatty acid degradation pathways suggests metabolic stress and a shift in energy sourcing. Notably, the upregulation of core glycolytic genes including hexokinase-1, fructose-bisphosphate aldolase, mitochondrial enolase superfamily member 1, and glyceraldehyde-3-phosphate dehydrogenase supports enhanced glycolytic activity [52–54]. Additionally, the upregulation of glucose dehydrogenase (FAD, quinone) and 6-phosphogluconate dehydrogenase suggests increased glucose and sugar derivative flux through glycolysis and the pentose phosphate pathway, likely supporting ATP and NADPH production to maintain redox balance under LCY-induced oxidative stress [52–54].

Elevated expressions of mitochondrial enzymes such as 2-oxoglutarate dehydrogenase, succinate semialdehyde dehydrogenase, aldehyde dehydrogenase, and L-threonine 3-dehydrogenase suggest an attempt to preserve energy metabolism and redox balance through enhanced TCA cycle function and amino acid catabolism [53,55–58]. Concurrent upregulation of short/branched chain-specific acyl-CoA dehydrogenase and isovaleryl-CoA dehydrogenase implies increased fatty acid and branched-chain amino acid oxidation as an alternative energy source [59,60]. Additionally, enrichment in the fatty acid degradation pathway along with the upregulation of NPC intracellular cholesterol transporters 1 and 2, lipase 3, and pancreatic triacylglycerol lipase-like suggests mobilization of lipid reserves as part of a broader metabolic adjustment [59,60]. This coordinated activation of glycolysis, lipid, and amino acid catabolism likely reflects a flexible strategy to meet elevated energetic demands imposed by LCY exposure during larval development.

Furthermore, detoxification pathways appear strongly activated, as evidenced by pronounced upregulation of cytochrome P450 family members (*cyp6a14, cyp314A1, cyp6k1, cyp9e2*) and various oxidoreductases, including short-chain dehydrogenase/reductase and retinal dehydrogenase 1, reflecting intensified metabolic processing of LCY and oxidative byproducts [51,61]. However, the marked downregulation of UDP-glucuronosyltransferase 1–3 and components of the mitochondrial electron transport chain (cytochrome b-c1 complex subunit 9, NADH–cytochrome b5 reductase 1) indicates impaired phase II detoxification and compromised mitochondrial respiration, potentially reducing ATP generation and exacerbating oxidative stress [62]. The overexpression of peroxidase points to a heightened oxidative burden, as this enzyme plays a crucial role in scavenging reactive oxygen species (ROS) generated during metabolic stress [61,63]. These observations were further supported by biochemical evidence showing reduced ATP production, increased lipid peroxidation (elevated MDA levels), and elevated oxidative stress markers in LCY-exposed larvae [11,50]. Taken together, our DEG data indicate that LCY exposure triggers a state of metabolic reprogramming in honey bee larvae, involving coordinated activation of glycolysis, the pentose phosphate pathway, amino acid catabolism, and lipid β-oxidation to meet increased energy and redox demands under oxidative stress. However, despite these adaptive responses, persistent mitochondrial dysfunction, impaired phase II detoxification, and elevated oxidative burden suggest that the cumulative stress exceeds the larvae's metabolic capacity, ultimately leading to adverse developmental outcomes such as reduced pupation success, lower adult emergence rates, and increased morphological deformities [11,32].

Thus, oxidative stress likely represents one of the key mechanisms of LCY toxicity. This is supported by KEGG enrichment in the ascorbate and aldarate metabolism pathway and GO terms such as oxidoreductase activity and FMN binding, all of which are involved in ROS metabolism and detoxification [61]. Notably, an 11-fold induction of peroxidase, a key antioxidant enzyme, indicates an acute oxidative response in LCY-exposed larvae, likely aimed at neutralizing excess hydrogen peroxide and related ROS [61]. This response is further supported by the upregulation of glucose oxidase and hydroxyacid oxidase 1, both of which generate or metabolize ROS as part of redox homeostasis [61]. Additionally, the increased expression of pyridoxine/pyridoxamine 5′-phosphate oxidase and pyridoxine-5′-phosphate oxidase, both essential for vitamin B6 metabolism, suggests an enhanced requirement for pyridoxal phosphate, a coenzyme involved in antioxidant processes and amino acid metabolism [61,64]. Despite these apparent protective mechanisms, LCY exposure is associated with the downregulation of mitochondrial RNA pseudouridine synthase and DNAJ homolog subfamily C member 18-like, implying mitochondrial dysfunction and impaired protein-folding capacity, which could compromise ATP production and activate apoptosis pathways [11,32,65]. Thus, while transcriptional upregulation of redox-related genes reflects an effort to mitigate LCY-induced oxidative damage, it may be insufficient to fully prevent mitochondrial impairment and cell death. These findings warrant further investigation into the balance between protective antioxidant responses and irreversible oxidative damage under LCY exposure.

Additionally, LCY exposure significantly disrupts sulfur metabolism in honey bee larvae, as evidenced by enriched GO terms such as sulfate transmembrane transporter activity and oxidoreductase activity, along with differential expression of key sulfur-related genes. Sulfur metabolism is vital for protein folding, antioxidant defense, and enzyme activity, and its dysregulation can impair detoxification efficiency and exacerbate cellular stress [61,63]. The transcriptional changes upon LCY exposure suggest a dual response involving the compensatory upregulation of protective genes alongside the suppression of essential mitochondrial functions. Notably, the upregulation of sodium-independent sulfate anion transporter isoforms and bifunctional 3′-phosphoadenosine 5′-phosphosulfate synthase 2 indicates an attempt to increase sulfate import and activation to meet the heightened sulfur demand during toxic stress [63]. The induction of homocysteine S-methyltransferase YbgG and 3-ketoacyl-CoA thiolase further supports an adaptive metabolic response aimed at maintaining methylation balance and enhancing lipid catabolism [63]. Despite these compensatory responses, the observed molecular disruptions suggest that larvae may have a limited capacity to detoxify LCY effectively [11]. However, these potentially protective mechanisms are undermined by the downregulation of key mitochondrial sulfur-handling and redox-related proteins, including iron-sulfur cluster co-chaperone protein HscB, S-adenosylmethionine mitochondrial carrier protein, and CDGSH iron-sulfur domain-containing protein 3, which are vital for cofactor biogenesis, electron transport, and oxidative phosphorylation [56,57]. Collectively, these changes indicate that although LCY-exposed larvae activate sulfur-based detoxification and metabolic responses, concurrent disruption of mitochondrial sulfur handling and redox balance likely limits their effectiveness, leading to persistent stress and overall reduction in detoxification capacity.

Neurotoxicity due to LCY exposure was evident through disruptions in the metabolism of tyrosine, tryptophan, and phenylalanine, which are essential for neurotransmitter biosynthesis [66]. These changes potentially alter dopamine and serotonin production, thereby affecting motor control, learning, and stress responses and likely contributing to the observed impairments in learning, olfactory memory, and foraging efficiency [14,15,31,66,67]. Such behavioral disruptions, if persistent, can potentially contribute to colony collapse [68,69]. Notably, the downregulation of leucine-rich repeat transmembrane neuronal protein 4-like and glutamate receptor ionotropic delta 2 suggests impaired excitatory signaling and reduced synaptic responsiveness [15,70]. Suppressed expression of acetylcholine receptor subunit alpha-10 and dopamine receptor D1 indicates further disruption of cholinergic and dopaminergic pathways crucial for cognitive and motor control [69,71–73]. Simultaneously, reduced expression of tropomyosin-2-like compromises neuronal structure and further impairs neuronal function [74]. Conversely, the upregulation of neurofilament heavy polypeptide, nicotinic acetylcholine receptor (alpha9), sensory neuron membrane protein 2, metabotropic glutamate receptor 7, leucine-rich repeat neuronal protein 1, and dopa mine N-acetyltransferase reflects compensatory efforts to preserve synaptic function and neurotransmitter balance [69,72,75–77]. The enrichment in neuropeptide and neuromedin U receptor activity further supports an adaptive response by larvae to counteract the synaptic dysfunction induced by LCY [78]. However, these responses may be inadequate to fully mitigate toxicity. Notably, behavioral impairments such as disorientation and abnormal foraging have also been reported in adult honey bees directly exposed to LCY, raising the possibility that similar molecular disruptions may underlie these outcomes despite differences in exposure stage [32,69].

Structural integrity appears compromised in LCY-exposed larvae, as evidenced by GO terms related to cuticle structure. Notably, the upregulation of cuticular proteins 13 and 14, which are integral to exoskeletal formation and rigidity [79,80], likely reflects a compensatory response to LCY-induced stress aimed at reinforcing the exoskeleton. However, despite this apparent molecular adaptation, LCY exposure results in malformed wings and antennae, along with reduced body size [11]. This discrepancy suggests that the protective upregulation of cuticular proteins may be insufficient to fully preserve structural integrity. Notably, the substantial downregulation of keratin-associated protein 19-2, along with collagen alpha-1(I) and collagen alpha-2(IV) chain-like, suggests a weakening of the extracellular scaffold necessary for tensile strength and elasticity of the cuticle [81,82].

These findings suggest that the downregulation of keratin-associated protein 19-2 and collagen alpha chains weakens the cuticle's supporting structure, likely reducing the effectiveness of the upregulated cuticular proteins. Nonetheless, this hypothesis requires further functional validation.

Overall, LCY exposure triggers widespread transcriptional changes in honey bee larvae, impacting neural activity, metabolic processes, structural development, and detoxification pathways. While several upregulated genes suggest the activation of compensatory responses, the overall pattern of dysregulation points to reduced physiological resilience and developmental fitness under LCY stress.

4.2. Persistent Neurobehavioral and Metabolic Dysfunctions in Adult Honey Bees Exposed to LCY

Chronic exposure to LCY during the larval stage appears to have lasting molecular effects that persist into adulthood, affecting key pathways related to neurobehavioral function and metabolism. The data highlight a complex interplay between disrupted neural signaling, impaired metabolic processes, and compromised detoxification mechanisms. These alterations may contribute to energy imbalances and cognitive vulnerabilities, potentially impacting individual fitness and overall colony stability [11,14,15,31,32,50].

Neurotoxicity may be a major and persistent toxic effect of LCY, particularly via the disruption of the neuroactive ligand–receptor interaction pathway (NLRP). Since the NLRP regulates synaptic transmission and neural communication, its disruption could impair learning and foraging behaviors in bees [31,32,83]. Furthermore, GO analysis highlighted the enrichment of G protein-coupled peptide receptor activity and peptide receptor activity, both of which play critical roles in neuromodulation, sensory perception, and hormone signaling, and are key components of the NLRP [84,85]. This suggests that LCY exposure not only disrupts neurotransmission but also alters stress and sensory response mechanisms.

Additionally, the upregulation of neuropeptides, including CAPA receptor-like protein, the nicotinic acetylcholine receptor (nAChR) alpha9 subunit, and the adipokinetic hormone receptor, suggests that LCY exposure alters neural signaling pathways, stress responses, and energy metabolism [71,84]. Notably, the upregulation of the nAChR alpha9 subunit, a key mediator of cholinergic neurotransmission, serves as a compensatory response to synaptic disruption but may heighten neural excitability and impair memory formation [11,71]. These molecular changes, indicative of disruptions in the NLRP, align with the observed learning deficits, as LCY-exposed bees exhibit poor performance in proboscis extension response learning assays [32]. Furthermore, the effect of LCY on peptide receptor activity may alter sensory processing and neurodevelopment, potentially explaining the disorientation and foraging inefficiencies observed in the exposed bees. This aligns with reports of a 90% decline in foraging activity following LCY exposure [15,31].

LCY exposure also disrupted tyrosine metabolism, a crucial pathway for neurotransmitter synthesis. The increased activity of homogentisate 1,2-dioxygenase, an enzyme essential for tyrosine breakdown, suggests either tyrosine depletion or a shift in metabolic flux. Furthermore, reduced aromatic-L-amino-acid decarboxylase activity indicates decreased biogenic amine production, which may exacerbate synaptic transmission deficits caused by LCY's direct impact on the NLRP. KEGG pathway enrichment analysis confirms that LCY specifically targets tyrosine metabolism and may reduce the availability of neurotransmitter precursors [66,86].

These metabolic disturbances, combined with LCY's direct effects on neuropeptide and acetylcholine receptors, likely contribute to the observed 90% decline in foraging activity [31] by impairing sensory processing, neuromodulation, and energy metabolism Thus, LCY-induced neurotoxicity appears to result from both direct interference with neural receptors and indirect disruption of neurotransmitter balance via altered tyrosine metabolism. These pathways are also closely linked to mitochondrial function, ATP production, and detoxification processes, highlighting oxidative stress as a central mechanism in LCY toxicity [54,58,87]. The upregulation of cytochrome b5, glucose dehydrogenase (FAD, quinone), and homogentisate 1,2-dioxygenase suggests that bees activate antioxidant defense mechanisms to mitigate ROS accumulation [87]. However, the downregulation of NADH dehydrogenase subunit 6 and NADH-cytochrome b5 reductase 1 indicates mitochondrial dysfunction, energy deficit, and impaired detoxification capacity, which may contribute to premature senescence and colony collapse [15,87]. These findings align with biochemical evidence of elevated lipid peroxidation and reduced antioxidant enzyme activities following LCY exposure [11,15,50].

Metabolic disruptions, particularly in lipid biosynthesis and transport, may further exacerbate LCY-induced neurotoxicity. GO analysis highlighted the enrichment in monoamine transmembrane transporter activity, which is essential for regulating dopamine and serotonin levels [88,89]. These neurotransmitters play critical roles in behavioral responses, stress adaptation, reproduction, and cognitive function [89–91]. The downregulation of NAD(P)H and monoamine-related transport genes suggests potential impairments in neurotransmitter metabolism, which may contribute to mood dysregulation, abnormal locomotion, and reduced foraging motivation [15,31,32]. Additionally, the observed enrichment of NAD(P)H activity is noteworthy because this enzyme is integral to DNA methylation and epigenetic regulation [92]. This suggests that LCY exposure influences gene expression beyond immediate physiological responses, potentially leading to long-term molecular and behavioral consequences [11,32,50].

The enrichment of the biosynthesis of unsaturated fatty acids pathway in LAG provides further insights into LCY-induced metabolic disruption. Downregulation of the elongation of very long-chain fatty acids protein 1 and acyl-CoA DeLCA(11) desaturase variant X3, alongside the upregulation of acyl-CoA DeLCA(11) desaturase variant X1, suggests impairments in membrane integrity, lipid storage, and stress resilience [15,93,94]. As unsaturated fatty acids are essential for cellular functions, hormone signaling, and immune defense, their disruption may weaken colony health, increase pathogen susceptibility, and delay development and behavior [11,15]. Additional metabolic disturbances were evident in enriched GO terms related to energy metabolism (FAD binding, heme binding, and oxidoreductase activity) and neurotransmitter function (monoamine transmembrane transporter activity) in LAG. Mitochondrial dysfunction, reflected in alterations to electron transport chain components, can directly affect fatty acid synthesis and vice versa [58]. Moreover, changes in lipid composition may disrupt the function of membrane-bound receptors, including G protein coupled peptide receptors and nicotinic acetylcholine receptors, potentially exacerbating LCY's neurotoxic effects [95]. Ultimately, the disruption of unsaturated fatty acid biosynthesis is a key factor linking metabolic dysfunction, oxidative stress, and neurotoxicity in LAG. Thus, the observed deficits in foraging activity, learning and memory, pollen and nectar storage, gut microbiota composition, and detoxification processes may collectively result from these widespread molecular disruptions [11,15,31,32].

Alterations in the detoxification pathways highlighted the limited ability of honey bees to metabolize LCY. The upregulation of cytochrome P450 9e2 and UDP-glucuronosyltransferase 2B17-like suggests an attempt to clear pesticide residues, whereas the downregulation of cytochrome P450 305a1 implies reduced LCY clearance efficiency [11,50]. As pyrethroids accumulate in tissues and disrupt multiple organ systems, these findings align with the systemic effects observed in adults exposed to LCY [96,97]. These molecular disruptions correspond with colony decline, reduced nectar collection, and weakened brood development, indicating prolonged pesticide retention and toxicity [15].

Developmental impairments in LCY-exposed honey bees suggest persistent structural and physiological deficiencies [11]. One notable alteration was the upregulation of cuticular protein 4, indicating potential disruption in exoskeleton formation, possibly due to LCY-induced cuticle damage or developmental stress [98]. Simultaneously, the downregulation of kinesin-like proteins (KIF9, KIN-7L, and K5A) suggests impairment of intracellular transport and structural maintenance, as these motor proteins are essential for cellular organization and function [99]. As intracellular transport is critical for proper cell growth, differentiation, and tissue integrity, such disruptions may contribute to the developmental abnormalities reported in LCY-exposed bees [11,99].

Endocrine-related disruptions in LAG were evident through the downregulation of juvenile hormone acid O-methyltransferase and methyl farnesoate epoxidase, suggesting dysregulation of hormonal signaling [100]. As these hormones regulate caste differentiation, reproduction, and colony organization, their disruption may lead to developmental impairments such as reduced pupation success, delayed emergence, and shortened adult lifespans [11,15]. Such effects may weaken colony stability and reproductive success, thereby contributing to population decline. Additionally, compromised immune and stress response pathways may increase susceptibility to environmental challenges. The upregulation of TOX high mobility group box family member 3-like suggests an adaptive immune activation response to LCY-induced physiological stress [101]. However, downregulation of the DNA repair protein RAD51 homolog 4 raises concerns about genomic instability, as impaired DNA repair capacity may increase vulnerability to DNA damage and long-term genetic impairments [102]. These interconnected molecular disruptions collectively threaten the health and resilience of honey bee colonies.

This study primarily presents transcriptomic analysis with qRT-PCR validation of selected DEGs, providing a solid foundation for understanding the molecular responses of honey bee larvae and adults to sublethal larval LCY exposure. However, transcriptome data inherently reflect gene expression changes at the RNA level and may not fully correspond to protein abundance or activity. Therefore, additional functional assays such as protein-level validation, enzymatic activity measurements, and gene-function studies are needed to further elucidate the mechanisms underlying LCY-induced phenotypic effects. Despite these limitations, the dataset offers valuable insights and serves as a critical resource for guiding future mechanistic research.

5. Conclusions

This study revealed significant transcriptomic disruptions in honey bee larvae and adults following LCY exposure, affecting key biological processes such as energy metabolism, neurotransmitter biosynthesis, oxidative stress responses, and detoxification pathways. These molecular alterations impair larval development, reduce pupation success, and compromise adult fitness, highlighting the systemic effects of LCY on honey bee physiology. Disruption of neuroactive ligand–receptor interactions and tyrosine metabolism suggests a potential link to the observed deficits in learning, foraging efficiency, and behavioral responses, while metabolic stress contributes to energy imbalances and oxidative damage. Structural deficiencies, such as weakened exoskeleton formation, may compromise survival. The limited detoxification capacity of honey bees underscores their vulnerability to LCY, as impaired enzymatic defenses appear insufficient to effectively mitigate toxicity. These findings provide molecular insights into the mechanisms underlying previously reported biochemical and behavioral impairments. Understanding these disruptions is crucial for refining pesticide risk assessments, developing targeted mitigation strategies, and promoting sustainable pest management practices to protect honey bee populations and maintain the ecological balance.

Supplementary Materials: The following supporting information can be downloaded at: https://www.mdpi.com/article/10.3390/insects16080833/s1, Figure S1: Schematic representation of the experimental workflow for LCY exposure and transcriptome analysis in *Apis mellifera*. First-instar larvae were grafted into 48-well plates (36 larvae per replicate) and reared at 35 °C and 95% relative humidity. From Day 3 to Day 6, larvae were chronically exposed to LCY (0.004 µg/larva; $n = 3$) or acetone control (0.5%; $n = 3$). Samples were collected at the larval stage (SLG: solvent-treated larvae group; LLG: LCY-treated larvae group; Day 8; three larvae per replicate) and adult stage (SAG: solvent-treated adult group; LAG: LCY-treated adult group; Day 19; three bees per replicate). Total RNA was extracted (RIN > 7), followed by cDNA library construction, Illumina paired-end sequencing (2 × 100 bp), read alignment to the *Apis mellifera* reference genome (*Amel_HAv3.1*), differential gene expression (DEG) analysis using DESeq2, Gene Ontology (GO) and Kyoto Encyclopedia of Genes and Genomes (KEGG) annotation, and qRT-PCR validation of selected DEGs. Figure S2: Comparative transcriptomic analyses of honey bee responses to lambda-cyhalothrin in honey bee larvae and adults. The heat map illustrates the differential expression of genes between four treatment comparisons: (A) SAG vs. SLG, (B) LAG vs. LLG, (C) LLG vs. SLG, and (D) LAG vs. SAG. The data are presented as z-scores of normalized log2-transformed expression values. Two-way hierarchical clustering was applied to organize the data. Gene expression differences are considered significant at a fold change of 1.5 with significance threshold *p*-values applied. SLG, solvent-treated larvae group; SAG, solvent-treated adult group; LLG, LCY-treated larvae group; LAG, LCY-treated adult group. Figure S3: Comparative transcriptomic analyses of honey bee responses to lambda-cyhalothrin in honey bee larvae and adults. Volcano plots illustrate the differential expression of genes between four treatment comparisons: (A) SAG vs. SLG, (B) LAG vs. LLG, (C) LLG vs. SLG, and (D) LAG vs. SAG. The data are presented as log2 fold changes against −log10 raw *p*-values. Gene expression differences are considered significant at a fold change (FC) of 1.5 and a raw p-value < 0.05, with upregulated genes shown in yellow and downregulated genes in blue. SLG, solvent-treated larvae group; SAG, solvent-treated adult group; LLG, LCY-treated larvae group; LAG, LCY-treated adult group. Figure S4: Top 10 cellular component terms of Gene Ontology (GO) functional analysis. (A) SAG vs. SLG and (B) LAG vs. LLG. Each bar represents the intersection size, with significance levels denoted by asterisks: a single asterisk (*) for *p*-values less than 0.05, double asterisks (**) for *p*-values less than 0.01, and triple asterisks (***) for *p*-values less than 0.001. SLG, solvent-treated larvae group; SAG, solvent-treated adult group; LLG, LCY treated larvae group; LAG, LCY-treated adult group. Figure S5: Top 10 molecular function terms of Gene Ontology (GO) functional analysis. (A) SAG vs. SLG and (B) LAG vs. LLG. Each bar represents the intersection size, with significance levels denoted by asterisks: a single asterisk (*) for *p*-values less than 0.05, double asterisks (**) for *p*-values less than 0.01, and triple asterisks (***) for *p*-values less than 0.001. SLG, solvent-treated larvae group; SAG, solvent-treated adult group; LLG, LCY-treated larvae group; LAG, LCY-treated adult group. Figure S6: Top 10 biological processes terms of Gene Ontology (GO) functional analysis. (A) SAG vs. SLG and (B) LAG vs. LLG. Each bar represents the intersection size, with significance levels denoted by asterisks: a single asterisk (*) for *p*-values less than 0.05, double asterisks (**) for *p*-values less than 0.01, and triple asterisks (***) for *p*-values less than 0.001. SLG, solvent-treated larvae group; SAG, solvent-treated adult group; LLG, LCY-treated larvae group; LAG, LCY-treated adult group. Figure S7: Comparative transcriptomic analysis of honey bee biological responses to lambda-cyhalothrin in larvae and adults. The heatmap shows the enrichment of various biological pathways across four treatment comparisons: SAG vs. SLG, LAG vs. LLG, LLG vs. SLG, and LAG vs. SAG. Pathways are categorized into different biological processes, indicated by different colors: Metabolism (red), Genetic Information Processing (orange), Environmental Information Processing (green), Cellular Processing (blue), Organismal Systems (purple), and Human Diseases (pink). The intensity of the blue color in each cell represents the significance level (*p*-value) of the enrichment test, with darker shades indicating more significant enrichment ($p \leq 0.001$) and lighter shades indicating less significant enrichment ($p > 0.05$). SLG, solvent-treated larvae group; SAG, solvent-treated adult group; LLG, LCY-treated larvae group; LAG, LCY-treated adult group. Figure S8: Top 10 KEGG pathways in different transcriptome comparisons. (A) SAG vs. SLG and (B) LAG vs. LLG. SLG, solvent-treated larvae group; SAG, solvent-treated adult group; LLG, LCY-treated larvae group; LAG, LCY-treated adult group. * $p < 0.05$, ** $p < 0.01$, *** $p < 0.001$. Figure S9: Correlation between transcriptome and qRT-PCR fold changes. A scatter plot showing the relationship between gene expression fold changes obtained from RNA-Seq (x-axis) and qRT-PCR (y-axis) analyses. The dotted line represents the linear regression fit ($R^2 = 0.6576$). Table S1: Oligo primers used for the validation of transcriptome data. Table S2: List of DEGs for GO_MF category in LLG compared to SLG. Table S3: List of DEGs for GO_MF category in LAG compared to SAG. Table S4: List of DEGs for GO_CC category in LAG compared to LLG. Table S5: List of DEGs for GO_CC category in SAG compared to SLG. Table S6: List of DEGs for GO_MF category in LAG compared to LLG. Table S7: List of DEGs for GO_MF category in SAG compared to SLG. Table S8: List of DEGs for GO_BP category in LAG compared to LLG. Table S9: List of DEGs for GO_BP category in SAG compared to SLG. Table S10: List of significant DEGs for each enriched KEGG pathway in LLG compared to SLG. Table S11: List of significant DEGs for each enriched KEGG pathway in LAG compared to SAG. Table S12: List of significant DEGs for each enriched KEGG pathway in SAG compared to SLG. Table S13: List of significant DEGs for each enriched KEGG pathway in LAG compared to LLG. Table S14: Summary of qRT-PCR validation results for selected differentially expressed genes across treatment comparisons. File S1: Comprehensive list of differentially expressed genes (DEGs) identified across all comparisons of honey bee samples at a fold change (FC) ≥ 1.5 and raw p-value < 0.05, including SAG vs. SLG, LAG vs. LLG, LLG vs. SLG, and LAG vs. SAG; File S2: Detailed list of shared and common condition-specific DEGs following LCY exposure.

Author Contributions: Conceptualization, B.M.K.V. and K.C.; methodology, B.M.K.V. and K.C.; software—validation, B.M.K.V., K.-H.P. and K.C.; formal analysis, B.M.K.V. and J.K.; investigation, B.M.K.V., J.K., C.-Y.Y., S.H. and M.C.; resources—data curation, writing—original draft preparation, B.M.K.V.; writing—review and editing, K.C. and B.M.K.V.; visualization, B.M.K.V. and B.-S.K.; supervision, K.C. and K.-H.P.; project administration, K.C.; funding acquisition, K.C. All authors have read and agreed to the published version of the manuscript.

Funding: This work was supported by the Research Program for Agricultural Science and Technology Development [Project No. PJ01579701], National Institute of Agricultural Sciences, Rural Development Administration, Republic of Korea. It was also partly supported by the 2025 the RDA Fellowship Program of the National Institute of Agricultural Sciences, Rural Development Administration, Republic of Korea. The funders did not influence the study design, data collection and analysis, decision to publish, or preparation of the manuscript.

Data Availability Statement: The datasets generated and/or analyzed in this study are included in this paper and provided as Supplementary Information. Raw sequencing files used in this analysis are available upon request from the corresponding author.

Conflicts of Interest: The authors declare no conflicts of interest.

Abbreviations

The following abbreviations are used in this manuscript:

(NAD(P)H)	Nicotinamide Adenine Dinucleotide Phosphate
a.i.	Active Ingredient
acyl-CoA DeLCA	Acyl-Coenzyme A Dehydrogenase, Long Chain
akhr	Adipokine tic Hormone Receptor
ATP	Adenosine Triphosphate
BP	Biological Processes
CC	Cellular Components
cDNA	Complementary DNA

cpr	Cytochrome P450 Reductase
cyp	Cytochrome P450
Cyp314a1	Cytochrome P450 314a1
cyp314A1	Cytochrome P450 314A1
cyp6a14	Cytochrome P450 6a14
cyp6k1	Cytochrome P450 6k1
cyp9e2	Cytochrome P450 9e2
DEGs	Differentially Expressed Genes
DNA	Deoxyribonucleic Acid
dop1	Dopamine Receptor 1
ELISA	Enzyme Linked Immunosorbent Assay
FAD	Flavin Adenine Dinucleotide
FC	Fold Change
FMN	Flavin Mononucleotide
FPKM	Fragments Per Kilobase of transcript per Million mapped reads
gapdh	Glyceraldehyde-3-Phosphate Dehydrogenase
GC	Guanine and Cytosine
GO	Gene Ontology
HISAT	Hierarchical Indexing Spliced Alignment of Transcripts
hsp70	Heat Shock Protein 70
K5A	Kinesin Family Member 5A
KEGG	Kyoto Encyclopedia of Genes and Genomes
KIF9	Kinesin Family Member 9
KIN-7L	Kinesin-like protein KIN-7L
LAG	LCY-treated Adult Group
LCY	Lambda-cyhalothrin
LD50	Median Lethal Dose
LLG	LCY-treated Larvae Group
LOC100578129	Kinesin-like protein KIN-7I, transcript variant X2
LOC100578936	Methylenetetrahydrofolate reductase, transcript variant X1
LOC408452	Cytochrome P450 9e2
LOC408559	Retinal dehydrogenase 1
LOC408650	Inositol oxygenase
LOC410507	Sodium-independent sulfate anion transporter, transcript variant X1
LOC411140	Putative aldehyde dehydrogenase family 7 member A1 homolog
LOC411672	Neuropeptides capa receptor-like protein
LOC413816	Sodium-independent sulfate anion transporter, transcript variant X2
LOC551044	Glucose dehydrogenase [FAD, quinone], transcript variant X2
LOC551109	Kinesin 5A, transcript variant X1
LOC551179	Methyl farnesoate epoxidase
LOC551465	Homogentisate 1,2-dioxygenase
LOC551837	Long-chain-fatty-acid–CoA ligase ACSBG2, transcript variant X1
LOC727166	Acyl-CoA Delta(11) desaturase, transcript variant X1
map2	Microtubule Associated Protein 2
MDA	Malondialdehyde
MF	Molecular Functions
nAChR	Nicotinic Acetylcholine Receptor
nAChRa9	Nicotinic Acetylcholine Receptor Alpha 9
NCBI	National Center for Biotechnology Information
NLRP	Neuroactive Ligand–Receptor Interaction Pathway
nos	Nitric Oxide Synthase
NPC	Niemann-Pick C
Q20	Quality Score of 20

Q30	Quality Score of 30
qRT-PCR	Quantitative Real-time Polymerase Chain Reaction
RIN	RNA Integrity Number
Rpl13a	Ribosomal Protein L13a
rps5	Ribosomal Protein S5
SAG	Solvent-treated Adult Group
sifr	SIFamide Receptor
SLG	Solvent-treated Larvae Group
Sog	Short Gastrulation
TOX	Thymocyte Selection Associated High Mobility Group Box Protein
tpi	Triosephosphate Isomerase
TPM	Transcript Per Million
UDP	Uridine Diphosphate

References

1. Sillman, J.; Uusitalo, V.; Tapanen, T.; Salonen, A.; Soukka, R.; Kahiluoto, H. Contribution of honeybees towards the net environmental benefits of food. *Sci. Total Environ.* **2021**, *756*, 143880. [CrossRef]
2. Smith, K.M.; Loh, E.H.; Rostal, M.K.; Zambrana-Torrelio, C.M.; Mendiola, L.; Daszak, P. Pathogens, pests, and economics: Drivers of honey bee colony declines and losses. *EcoHealth* **2013**, *10*, 434–445. [CrossRef]
3. VanEngelsdorp, D.; Evans, J.D.; Saegerman, C.; Mullin, C.; Haubruge, E.; Nguyen, B.K.; Frazier, M.; Frazier, J.; Cox-Foster, D.; Chen, Y. Colony collapse disorder: A descriptive study. *PLoS ONE* **2009**, *4*, e6481. [CrossRef] [PubMed]
4. Sanchez-Bayo, F.; Goka, K. Pesticide residues and bees—A risk assessment. *PLoS ONE* **2014**, *9*, e94482. [CrossRef] [PubMed]
5. Zhu, W.; Schmehl, D.R.; Mullin, C.A.; Frazier, J.L. Four common pesticides, their mixtures and a formulation solvent in the hive environment have high oral toxicity to honey bee larvae. *PLoS ONE* **2014**, *9*, e77547. [CrossRef] [PubMed]
6. Chmiel, J.A.; Daisley, B.A.; Pitek, A.P.; Thompson, G.J.; Reid, G. Understanding the effects of sublethal pesticide exposure on honey bees: A role for probiotics as mediators of environmental stress. *Front. Ecol. Evol.* **2020**, *8*, 22. [CrossRef]
7. Krupke, C.H.; Hunt, G.J.; Eitzer, B.D.; Andino, G.; Given, K. Multiple routes of pesticide exposure for honey bees living near agricultural fields. *PLoS ONE* **2012**, *7*, e29268. [CrossRef]
8. Wueppenhorst, K.; Eckert, J.H.; Steinert, M.; Erler, S. What about honey bee jelly? Pesticide residues in larval food jelly of the Western honey bee *Apis mellifera*. *Sci. Total Environ.* **2022**, *850*, 158095. [CrossRef]
9. Wu, J.Y.; Anelli, C.M.; Sheppard, W.S. Sub-lethal effects of pesticide residues in brood comb on worker honey bee (*Apis mellifera*) development and longevity. *PLoS ONE* **2011**, *6*, e14720. [CrossRef]
10. Xiao, J.; He, Q.; Liu, Q.; Wang, Z.; Yin, F.; Chai, Y.; Yang, Q.; Jiang, X.; Liao, M.; Yu, L. Analysis of honey bee exposure to multiple pesticide residues in the hive environment. *Sci. Total Environ.* **2022**, *805*, 150292. [CrossRef]
11. Choi, J.-Y.; Chon, K.; Kim, J.; Vasamsetti, B.M.K.; Kim, B.-S.; Yoon, C.-Y.; Hwang, S.; Park, K.-H.; Lee, J.-H. Assessment of Lambda-Cyhalothrin and Spinetoram Toxicity and Their Effects on the Activities of Antioxidant Enzymes and Acetylcholinesterase in Honey Bee (*Apis mellifera*) Larvae. *Insects* **2024**, *15*, 587. [CrossRef]
12. Kim, J.; Chon, K.; Kim, B.S.; Oh, J.A.; Yoon, C.Y.; Park, H.H. Assessment of acute and chronic toxicity of cyantraniliprole and sulfoxaflor on honey bee (*Apis mellifera*) larvae. *Pest Manag. Sci.* **2022**, *78*, 5402–5412. [CrossRef] [PubMed]
13. Li, B.; Ke, L.; Li, A.-R.; Diao, Q.-Y.; Wang, Q.; Liu, Y.-J. Exposure of larvae to sublethal thiacloprid delays bee development and affects transcriptional responses of newly emerged honey bees. *Front. Insect Sci.* **2022**, *2*, 844957. [CrossRef]
14. Waseem, M.A.; Thakur, M.; Singh, M.P.; Kasi, I.K. Evaluation of lambda cyhalothrin toxicity to Indian honeybees Apis Cerana in laboratory conditions. *J. Entomol. Res.* **2024**, *48*, 214–219. [CrossRef]
15. Waseem, M.A.; Thakur, M.; Vallabuni, S.; Sharma, S.; Hashem, A.; Abd Allah, E.F. Toxicological Impact of Lambda-Cyhalothrin on *Apis mellifera*: Comparative Analysis Under Semi-Field and Field Conditions. *J. Apic. Sci.* **2024**, *68*, 119–131. [CrossRef]
16. Furlan, L.; Kreutzweiser, D. Alternatives to neonicotinoid insecticides for pest control: Case studies in agriculture and forestry. *Environ. Sci. Pollut. Res.* **2015**, *22*, 135–147. [CrossRef]
17. Grout, T.A.; Koenig, P.A.; Kapuvari, J.K.; McArt, S.H. Neonicotinoid insecticides in New York state: Economic benefits and risk to pollinators. *Renew. Resour. J.* **2021**, *35*, 18–22.
18. Jactel, H.; Verheggen, F.; Thiéry, D.; Escobar-Gutiérrez, A.J.; Gachet, E.; Desneux, N.; Group, N.W. Alternatives to neonicotinoids. *Environ. Int.* **2019**, *129*, 423–429. [CrossRef]
19. Liu, Q.; He, Q.; Zhang, S.; Chai, Y.; Gao, Q.; Xiao, J.; Fang, Q.; Yu, L.; Cao, H. Toxic effects of detected pyrethroid pesticides on honeybee (*Apis mellifera* ligustica Spin and Apis cerana cerana Fabricius). *Sci. Rep.* **2022**, *12*, 16695. [CrossRef]

20. Drew, W.T.; Parmar A, H.K.; Lambda-Cyhalothrin, D.M. Human Health Risk Assessment for the Proposed Food/Feed Uses of the Insecticide on Cucurbit Vegetables (Group 9), Tuberous and Corm Vegetables (Subgroup 1C), Grass Forage, Fodder, and Hay (Group 17), Barley, Buckwheat, Oat, Rye, Wild Rice, and Pistachios; Petition Numbers 5F6994, 3E6593, and 6E7077EPA. Office of Prevention, Pesticides, and Toxic Substances Risk Assessment. 2007. Available online: https://www3.epa.gov/pesticides/endanger/litstatus/effects/redleg-frog/2012/lambda-cyha/appendix-j.pdf (accessed on 25 April 2024).
21. Burr, S.A.; Ray, D.E. Structure-activity and interaction effects of 14 different pyrethroids on voltage-gated chloride ion channels. *Toxicol. Sci.* **2004**, *77*, 341–346. [CrossRef]
22. Naravaneni, R.; Jamil, K. Evaluation of cytogenetic effects of lambda-cyhalothrin on human lymphocytes. *J. Biochem. Mol. Toxicol.* **2005**, *19*, 304–310. [CrossRef] [PubMed]
23. Deeba, F.; Raza, I.; Muhammad, N.; Rahman, H.; ur Rehman, Z.; Azizullah, A.; Khattak, B.; Ullah, F.; Daud, M. Chlorpyrifos and lambda cyhalothrin-induced oxidative stress in human erythrocytes: In vitro studies. *Toxicol. Ind. Health* **2017**, *33*, 297–307. [CrossRef] [PubMed]
24. Hsu, S.-S.; Jan, C.-R.; Liang, W.-Z. The investigation of the pyrethroid insecticide lambda-cyhalothrin (LCT)-affected Ca^{2+} homeostasis and-activated Ca^{2+}-associated mitochondrial apoptotic pathway in normal human astrocytes: The evaluation of protective effects of BAPTA-AM (a selective Ca^{2+} chelator). *Neurotoxicology* **2018**, *69*, 97–107. [PubMed]
25. Fetoui, H.; Garoui, E.M.; Zeghal, N. Lambda-cyhalothrin-induced biochemical and histopathological changes in the liver of rats: Ameliorative effect of ascorbic acid. *Exp. Toxicol. Pathol.* **2009**, *61*, 189–196. [CrossRef]
26. Aouey, B.; Derbali, M.; Chtourou, Y.; Bouchard, M.; Khabir, A.; Fetoui, H. Pyrethroid insecticide lambda-cyhalothrin and its metabolites induce liver injury through the activation of oxidative stress and proinflammatory gene expression in rats following acute and subchronic exposure. *Environ. Sci. Pollut. Res.* **2017**, *24*, 5841–5856. [CrossRef]
27. Fetoui, H.; Makni, M.; Garoui, E.M.; Zeghal, N. Toxic effects of lambda-cyhalothrin, a synthetic pyrethroid pesticide, on the rat kidney: Involvement of oxidative stress and protective role of ascorbic acid. *Exp. Toxicol. Pathol.* **2010**, *62*, 593–599. [CrossRef]
28. Breckenridge, C.B.; Holden, L.; Sturgess, N.; Weiner, M.; Sheets, L.; Sargent, D.; Soderlund, D.M.; Choi, J.-S.; Symington, S.; Clark, J.M. Evidence for a separate mechanism of toxicity for the Type I and the Type II pyrethroid insecticides. *Neurotoxicology* **2009**, *30*, S17–S31. [CrossRef]
29. Tong, Z.; Duan, J.; Wu, Y.; Liu, Q.; He, Q.; Shi, Y.; Yu, L.; Cao, H. A survey of multiple pesticide residues in pollen and beebread collected in China. *Sci. Total Environ.* **2018**, *640*, 1578–1586. [CrossRef]
30. Kaila, L.; Ketola, J.; Toivonen, M.; Loukola, O.; Hakala, K.; Raiskio, S.; Hurme, T.; Jalli, M. Pesticide residues in honeybee-collected pollen: Does the EU regulation protect honeybees from pesticides? *Environ. Sci. Pollut. Res.* **2022**, *29*, 18225–18244. [CrossRef]
31. Waseem, M.A.; Sahoo, B.K.; Mohanty, P.; Thakur, M. Impact of Fipronil and Lambda-Cyhalothrin on Foraging Behaviour of *Apis dorsata* in Onion. *Indian J. Entomol.* **2024**, 1–5. [CrossRef]
32. Liao, C.-h.; He, X.-j.; Wang, Z.-l.; Barron, A.B.; Zhang, B.; Zeng, Z.-j.; Wu, X.-b. Short-term exposure to lambda-cyhalothrin negatively affects the survival and memory-related characteristics of worker bees *Apis mellifera*. *Arch. Environ. Contam. Toxicol.* **2018**, *75*, 59–65. [CrossRef] [PubMed]
33. Guo, D.; Li, Z.; Zhang, Y.; Zhang, W.; Wang, C.; Zhang, D.-x.; Liu, F.; Gao, Z.; Xu, B.; Wang, N. The effect of lambda-cyhalothrin nanocapsules on the gut microbial communities and immune response of the bee elucidates the potential environmental impact of emerging nanopesticides. *J. Hazard. Mater.* **2024**, *479*, 135650. [CrossRef] [PubMed]
34. Ingram, E.M.; Augustin, J.; Ellis, M.D.; Siegfried, B.D. Evaluating sub-lethal effects of orchard-applied pyrethroids using video-tracking software to quantify honey bee behaviors. *Chemosphere* **2015**, *135*, 272–277. [CrossRef] [PubMed]
35. Abdel Razik, M.A.R.A.M. Toxicity and side effects of some insecticides applied in cotton fields on *Apis mellifera*. *Environ. Sci. Pollut. Res.* **2019**, *26*, 4987–4996. [CrossRef]
36. PPDB: Pesticide Properties DataBase. Lambda-Cyhalothrin (Ref: OMS 3021). Available online: http://sitem.herts.ac.uk/aeru/ppdb/en/Reports/415.htm (accessed on 20 March 2024).
37. Johnson, R.M. Honey bee toxicology. *Annu. Rev. Entomol.* **2015**, *60*, 415–434. [CrossRef]
38. Vasamsetti, B.M.K.; Chon, K.; Yoon, C.-Y.; Kim, J.; Choi, J.-Y.; Hwang, S.; Park, K.-H. Transcriptome Profiling of Etridiazole-Exposed Zebrafish (*Danio rerio*) Embryos Reveals Pathways Associated with Cardiac and Ocular Toxicities. *Int. J. Mol. Sci.* **2023**, *24*, 15067. [CrossRef]
39. Vasamsetti, B.M.K.; Chon, K.; Choi, J.-Y.; Kim, J.; Yoon, C.-Y. Transcriptome Analysis of Thiram-Treated Zebrafish (*Danio rerio*) Embryos Reveals Disruption of Reproduction Signaling Pathways. *Biology* **2023**, *12*, 156. [CrossRef]
40. Vasamsetti, B.M.K.; Chon, K.; Kim, J.; Oh, J.-A.; Yoon, C.-Y.; Park, H.-H. Transcriptome-based identification of genes responding to the organophosphate pesticide phosmet in *Danio rerio*. *Genes* **2021**, *12*, 1738. [CrossRef]
41. Park, M.Y.; Krishna Vasamsetti, B.M.; Kim, W.S.; Kang, H.J.; Kim, D.-Y.; Lim, B.; Cho, K.; Kim, J.S.; Chee, H.K.; Park, J.H. Comprehensive analysis of cardiac xeno-graft unveils rejection mechanisms. *Int. J. Mol. Sci.* **2021**, *22*, 751. [CrossRef]
42. Vasamsetti, B.M.K.; Kim, J.; Chon, K.; Kim, B.-S.; Yoon, C.-Y.; Hwang, S.; Park, K.-H. Molecular Impact of Sublethal Spinetoram Exposure on Honeybee (*Apis mellifera*) Larval and Adult Transcriptomes. *Int. J. Mol. Sci.* **2024**, *25*, 11923. [CrossRef]

43. Bolger, A.M.; Lohse, M.; Usadel, B. Trimmomatic: A flexible trimmer for Illumina sequence data. *Bioinformatics* **2014**, *30*, 2114–2120. [CrossRef]
44. Kim, D.; Paggi, J.M.; Park, C.; Bennett, C.; Salzberg, S.L. Graph-based genome alignment and genotyping with HISAT2 and HISAT-genotype. *Nat. Biotechnol.* **2019**, *37*, 907–915. [CrossRef]
45. Li, H.; Handsaker, B.; Wysoker, A.; Fennell, T.; Ruan, J.; Homer, N.; Marth, G.; Abecasis, G.; Durbin, R.; Subgroup, G.P.D.P. The sequence alignment/map format and SAMtools. *Bioinformatics* **2009**, *25*, 2078–2079. [CrossRef] [PubMed]
46. Pertea, M.; Pertea, G.M.; Antonescu, C.M.; Chang, T.-C.; Mendell, J.T.; Salzberg, S.L. StringTie enables improved reconstruction of a transcriptome from RNA-seq reads. *Nat. Biotechnol.* **2015**, *33*, 290–295. [CrossRef]
47. Love, M.I.; Huber, W.; Anders, S. Moderated estimation of fold change and dispersion for RNA-seq data with DESeq2. *Genome Biol.* **2014**, *15*, 550. [CrossRef] [PubMed]
48. Raudvere, U.; Kolberg, L.; Kuzmin, I.; Arak, T.; Adler, P.; Peterson, H.; Vilo, J. g:Profiler: A web server for functional enrichment analysis and conversions of gene lists (2019 update). *Nucleic Acids Res.* **2019**, *47*, W191–W198. [CrossRef] [PubMed]
49. Livak, K.J.; Schmittgen, T.D. Analysis of relative gene expression data using real-time quantitative PCR and the $2-\Delta\Delta CT$ method. *Methods* **2001**, *25*, 402–408. [CrossRef]
50. Chen, X.; Wang, F.; Guo, H.; Liu, X.; Wu, S.; Lv, L.; Tang, T. Uncovering hidden dangers: The combined toxicity of abamectin and lambda-cyhalothrin on honey bees. *Sci. Total Environ.* **2024**, *933*, 173126. [CrossRef]
51. Berenbaum, M.R.; Johnson, R.M. Xenobiotic detoxification pathways in honey bees. *Curr. Opin. Insect Sci.* **2015**, *10*, 51–58. [CrossRef]
52. Nirupama, R.; Rajaraman, B.; Yajurvedi, H. Stress and glucose metabolism: A review. *Imaging J. Clin. Med. Sci.* **2018**, *5*, 008–012.
53. Ghani, M.U.; Yang, Z.; Feng, T.; Chen, J.; Khosravi, Z.; Wu, Q.; Cui, H. Comprehensive review on glucose 6 phosphate dehydrogenase: A critical immunometabolic and redox switch in insects. *Int. J. Biol. Macromol.* **2024**, *273*, 132867. [CrossRef]
54. Kunieda, T.; Fujiyuki, T.; Kucharski, R.; Foret, S.; Ament, S.; Toth, A.; Ohashi, K.; Takeuchi, H.; Kamikouchi, A.; Kage, E. Carbohydrate metabolism genes and pathways in insects: Insights from the honey bee genome. *Insect Mol. Biol.* **2006**, *15*, 563–576. [CrossRef]
55. Torres, N.; Tobón-Cornejo, S.; Velazquez-Villegas, L.A.; Noriega, L.G.; Alemán-Escondrillas, G.; Tovar, A.R. Amino acid catabolism: An overlooked area of metabolism. *Nutrients* **2023**, *15*, 3378. [CrossRef] [PubMed]
56. Handy, D.E.; Loscalzo, J. Redox regulation of mitochondrial function. *Antioxid. Redox Signal.* **2012**, *16*, 1323–1367. [CrossRef] [PubMed]
57. Puglisi, R.; Pastore, A. The role of chaperones in iron–sulfur cluster biogenesis. *FEBS Lett.* **2018**, *592*, 4011–4019. [CrossRef] [PubMed]
58. Wedan, R.J.; Longenecker, J.Z.; Nowinski, S.M. Mitochondrial fatty acid synthesis is an emergent central regulator of mammalian oxidative metabolism. *Cell Metab.* **2024**, *36*, 36–47. [CrossRef]
59. Furse, S.; Koch, H.; Wright, G.A.; Stevenson, P.C. Sterol and lipid metabolism in bees. *Metabolomics* **2023**, *19*, 78. [CrossRef]
60. Castaños, C.E.; Boyce, M.C.; Bates, T.; Millar, A.H.; Flematti, G.; Lawler, N.G.; Grassl, J. Lipidomic features of honey bee and colony health during limited supplementary feeding. *Insect Mol. Biol.* **2023**, *32*, 658–675. [CrossRef]
61. du Rand, E.d.; Smit, S.; Beukes, M.; Apostolides, Z.; Pirk, C.; Nicolson, S. Detoxification mechanisms of honey bees (*Apis mellifera*) resulting in tolerance of dietary nicotine. *Sci. Rep.* **2015**, *5*, 11779. [CrossRef]
62. Cui, X.; Wang, C.; Wang, X.; Li, G.; Liu, Z.; Wang, H.; Guo, X.; Xu, B. Molecular Mechanism of the UDP-Glucuronosyltransferase 2B20-like Gene (AccUGT2B20-like) in Pesticide Resistance of Apis cerana cerana. *Front. Genet.* **2020**, *11*, 592595. [CrossRef]
63. Yu, S.J. Detoxification Mechanisms in Insects. In *Encyclopedia of Entomology*; Capinera, J.L., Ed.; Springer: Dordrecht, The Netherlands, 2008; pp. 1187–1201.
64. Nazemi-Rafie, J.; Fatehi, F.; Hasrak, S. A comparative transcriptome analysis of the head of 1 and 9 days old worker honeybees (*Apis mellifera*). *Bull. Entomol. Res.* **2023**, *113*, 253–270. [CrossRef] [PubMed]
65. Antonicka, H.; Choquet, K.; Lin, Z.Y.; Gingras, A.C.; Kleinman, C.L.; Shoubridge, E.A. A pseudouridine synthase module is essential for mitochondrial protein synthesis and cell viability. *EMBO Rep.* **2017**, *18*, 28–38. [CrossRef] [PubMed]
66. Sasaki, K. Nutrition and dopamine: An intake of tyrosine in royal jelly can affect the brain levels of dopamine in male honeybees (*Apis mellifera* L.). *J. Insect Physiol.* **2016**, *87*, 45–52. [CrossRef]
67. Fan, W.; Li, G.; Zhang, X.; Wang, Y.; Wang, C.; Xu, B.; Guo, X.; Li, H. The role of melatonin and Tryptophan-5-hydroxylase-1 in different abiotic stressors in Apis cerana cerana. *J. Insect Physiol.* **2021**, *128*, 104180. [CrossRef]
68. Harris, J.W.; Woodring, J. Effects of stress, age, season, and source colony on levels of octopamine, dopamine and serotonin in the honey bee (*Apis mellifera* L.) brain. *J. Insect Physiol.* **1992**, *38*, 29–35. [CrossRef]
69. Belzunces, L.P.; Tchamitchian, S.; Brunet, J.-L. Neural effects of insecticides in the honey bee. *Apidologie* **2012**, *43*, 348–370. [CrossRef]
70. Schroeder, A.; de Wit, J. Leucine-rich repeat-containing synaptic adhesion molecules as organizers of synaptic specificity and diversity. *Exp. Mol. Med.* **2018**, *50*, 1–9. [CrossRef]

71. Jones, A.K.; Raymond-Delpech, V.; Thany, S.H.; Gauthier, M.; Sattelle, D.B. The nicotinic acetylcholine receptor gene family of the honey bee, *Apis mellifera*. *Genome Res.* **2006**, *16*, 1422–1430. [CrossRef]
72. Lu, W.; Liu, Z.; Fan, X.; Zhang, X.; Qiao, X.; Huang, J. Nicotinic acetylcholine receptor modulator insecticides act on diverse receptor subtypes with distinct subunit compositions. *PLoS Genet.* **2022**, *18*, e1009920. [CrossRef]
73. Mustard, J.A.; Pham, P.M.; Smith, B.H. Modulation of motor behavior by dopamine and the D1-like dopamine receptor AmDOP2 in the honey bee. *J. Insect Physiol.* **2010**, *56*, 422–430. [CrossRef]
74. Gray, K.T.; Kostyukova, A.S.; Fath, T. Actin regulation by tropomodulin and tropomyosin in neuronal morphogenesis and function. *Mol. Cell. Neurosci.* **2017**, *84*, 48–57. [CrossRef]
75. Kim, G.; Lujan, R.; Schwenk, J.; Kelley, M.H.; Aguado, C.; Watanabe, M.; Fakler, B.; Maylie, J.; Adelman, J.P. Membrane palmitoylated protein 2 is a synaptic scaffold protein required for synaptic SK2-containing channel function. *eLife* **2016**, *5*, e12637. [CrossRef]
76. Martín, R.; Durroux, T.; Ciruela, F.; Torres, M.; Pin, J.-P.; Sánchez-Prieto, J. The metabotropic glutamate receptor mGlu7 activates phospholipase C, translocates munc-13-1 protein, and potentiates glutamate release at cerebrocortical nerve terminals. *J. Biol. Chem.* **2010**, *285*, 17907–17917. [CrossRef]
77. Liu, B.; Zhang, Y.; Fan, Y.; Wang, S.; Li, Z.; Deng, M.; Li, C.; Wang, J.; Ma, R.; Wang, X. Leucine-rich repeat neuronal protein-1 suppresses apoptosis of gastric cancer cells through regulation of Fas/FasL. *Cancer Sci.* **2019**, *110*, 2145–2155. [CrossRef] [PubMed]
78. Urlacher, E.; Soustelle, L.; Parmentier, M.-L.; Verlinden, H.; Gherardi, M.-J.; Fourmy, D.; Mercer, A.R.; Devaud, J.-M.; Massou, I. Honey bee allatostatins target galanin/somatostatin-like receptors and modulate learning: A conserved function? *PLoS ONE* **2016**, *11*, e0146248. [CrossRef] [PubMed]
79. Zhao, X.; Gou, X.; Qin, Z.; Li, D.; Wang, Y.; Ma, E.; Li, S.; Zhang, J. Identification and expression of cuticular protein genes based on Locusta migratoria transcriptome. *Sci. Rep.* **2017**, *7*, 45462. [CrossRef]
80. Soares, M.P.; Elias-Neto, M.; Simões, Z.L.; Bitondi, M.M. A cuticle protein gene in the honeybee: Expression during development and in relation to the ecdysteroid titer. *Insect Biochem. Mol. Biol.* **2007**, *37*, 1272–1282. [CrossRef]
81. Ramms, L.; Fabris, G.; Windoffer, R.; Schwarz, N.; Springer, R.; Zhou, C.; Lazar, J.; Stiefel, S.; Hersch, N.; Schnakenberg, U. Keratins as the main component for the mechanical integrity of keratinocytes. *Proc. Natl. Acad. Sci. USA* **2013**, *110*, 18513–18518. [CrossRef]
82. Summers, J.A.; Yarbrough, M.; Liu, M.; McDonald, W.H.; Hudson, B.G.; Pastor-Pareja, J.C.; Boudko, S.P. Collagen IV of basement membranes: IV. Adaptive mechanism of collagen IV scaffold assembly in Drosophila. *J. Biol. Chem.* **2023**, *299*, 105394. [CrossRef]
83. Huang, J.; Wang, T.; Qiu, Y.; Hassanyar, A.K.; Zhang, Z.; Sun, Q.; Ni, X.; Yu, K.; Guo, Y.; Yang, C. Differential brain expression patterns of microRNAs related to olfactory performance in honey bees (*Apis mellifera*). *Genes* **2023**, *14*, 1000. [CrossRef]
84. Liu, N.; Wang, Y.; Li, T.; Feng, X. G-protein coupled receptors (GPCRs): Signaling pathways, characterization, and functions in insect physiology and toxicology. *Int. J. Mol. Sci.* **2021**, *22*, 5260. [CrossRef] [PubMed]
85. Liao, J.; Wan, K.; Lü, Y.; Ouyang, W.; Huang, J.; Zheng, L.; Miao, L.; Su, S.; Li, Z. Comparison of Brain Gene Expression Profiles Associated with Auto-Grooming Behavior between Apis cerana and *Apis mellifera* Infested by Varroa destructor. *Genes* **2024**, *15*, 763. [CrossRef] [PubMed]
86. Wang, B.; Habermehl, C.; Jiang, L. Metabolomic analysis of honey bee (*Apis mellifera* L.) response to glyphosate exposure. *Mol. Omics* **2022**, *18*, 635–642. [CrossRef] [PubMed]
87. Olgun, T.; Dayıoğlu, M.; Özsoy, N. Pesticide and pathogen induced oxidative stress in honey bees (*Apis mellifera* L.). *Mellifera* **2020**, *20*, 32–52.
88. Zhang, V.; Kucharski, R.; Landers, C.; Richards, S.N.; Bröer, S.; Martin, R.E.; Maleszka, R. Characterization of a dopamine transporter and its splice variant reveals novel features of dopaminergic regulation in the honey bee. *Front. Physiol.* **2019**, *10*, 1375. [CrossRef]
89. Wright, G. The role of dopamine and serotonin in conditioned food aversion learning in the honeybee. *Commun. Integr. Biol.* **2011**, *4*, 318–320. [CrossRef]
90. Huang, Z.-Y.; Robinson, G.E.; Tobe, S.S.; Yagi, K.J.; Strambi, C.; Strambi, A.; Stay, B. Hormonal regulation of behavioural development in the honey bee is based on changes in the rate of juvenile hormone biosynthesis. *J. Insect Physiol.* **1991**, *37*, 733–741. [CrossRef]
91. Robinson, G.E.; Strambi, C.; Strambi, A.; Huang, Z.-Y. Reproduction in worker honey bees is associated with low juvenile hormone titers and rates of biosynthesis. *Gen. Comp. Endocrinol.* **1992**, *87*, 471–480. [CrossRef]
92. Friso, S.; Choi, S.-W.; Girelli, D.; Mason, J.B.; Dolnikowski, G.G.; Bagley, P.J.; Olivieri, O.; Jacques, P.F.; Rosenberg, I.H.; Corrocher, R. A common mutation in the 5,10-methylenetetrahydrofolate reductase gene affects genomic DNA methylation through an interaction with folate status. *Proc. Natl. Acad. Sci. USA* **2002**, *99*, 5606–5611. [CrossRef]
93. Cerone, M.; Smith, T.K. Desaturases: Structural and mechanistic insights into the biosynthesis of unsaturated fatty acids. *Iubmb Life* **2022**, *74*, 1036–1051. [CrossRef]

94. Wang, X.; Yu, H.; Gao, R.; Liu, M.; Xie, W. A comprehensive review of the family of very-long-chain fatty acid elongases: Structure, function, and implications in physiology and pathology. *Eur. J. Med. Res.* **2023**, *28*, 532. [CrossRef]
95. Fang, Y.; Feng, M.; Ma, C.; Rueppell, O.; Li, J. Major royal jelly proteins influence the neurobiological regulation of the division of labor among honey bee workers. *Int. J. Biol. Macromol.* **2023**, *225*, 848–860. [CrossRef]
96. Anadon, A.; Martinez, M.; Martinez, M.; Diaz, M.; Martinez-Larranaga, M. Toxicokinetics of lambda-cyhalothrin in rats. *Toxicol. Lett.* **2006**, *165*, 47–56. [CrossRef] [PubMed]
97. Rehman, H.; Aziz, A.-T.; Saggu, S.; Abbas, Z.K.; Mohan, A.; Ansari, A.A. Systematic review on pyrethroid toxicity with special reference to deltamethrin. *J. Entomol. Zool. Stud.* **2014**, *2*, 60–70.
98. Falcón, T.; Ferreira-Caliman, M.J.; Nunes, F.M.F.; Tanaka, É.D.; do Nascimento, F.S.; Bitondi, M.M.G. Exoskeleton formation in *Apis mellifera*: Cuticular hydrocarbons profiles and expression of desaturase and elongase genes during pupal and adult development. *Insect Biochem. Mol. Biol.* **2014**, *50*, 68–81. [CrossRef] [PubMed]
99. Hirokawa, N.; Noda, Y.; Tanaka, Y.; Niwa, S. Kinesin superfamily motor proteins and intracellular transport. *Nat. Rev. Mol. Cell Biol.* **2009**, *10*, 682–696. [CrossRef]
100. Smykal, V.; Dolezel, D. Evolution of proteins involved in the final steps of juvenile hormone synthesis. *J. Insect Physiol.* **2023**, *145*, 104487. [CrossRef]
101. Maestre, L.; García-García, J.F.; Jiménez, S.; Reyes-García, A.I.; García-González, Á.; Montes-Moreno, S.; Arribas, A.J.; González-García, P.; Caleiras, E.; Banham, A.H. High-mobility group box (TOX) antibody a useful tool for the identification of B and T cell subpopulations. *PLoS ONE* **2020**, *15*, e0229743. [CrossRef]
102. Fouks, B.; Miller, K.J.; Ross, C.; Jones, C.; Rueppell, O. Alternative double strand break repair pathways shape the evolution of high recombination in the honey bee, *Apis mellifera*. *Insect Mol. Biol.* **2025**, *34*, 185–202. [CrossRef]

insects

Article

Impact of Atrazine on Sucrose Sensitivity in Honey Bees

Xiexin Hu [1,2], Zixuan Xu [2], Jiachen Xu [2], Guiyi Ma [2], Yiren Pan [2]Minqi Cai [2], Zheguang Lin [2], Ting Ji [2] and Kang Wang [1,2,*]

1 College of Bioscience and Biotechnology, Yangzhou University, Yangzhou 225009, China; huxiexin@outlook.com
2 College of Animal Science and Technology, Yangzhou University, Yangzhou 225009, China
* Correspondence: kwang@yzu.edu.cn

Simple Summary: Honey bees play a crucial role in pollinating crops and maintaining biodiversity; however, their populations are declining due to various environmental threats, including pesticide exposure. In this study, we investigated the effects of atrazine, a widely used herbicide, on honey bee behavior and brain function. Specifically, we tested whether atrazine exposure influences the ability of bees to sense sweetness, which is essential for finding food. Our results showed that bees exposed to atrazine had a reduced ability to detect sucrose solution, potentially impairing their foraging efficiency. A transcriptomic analysis revealed significant changes in gene expression levels in the brain, particularly in genes linked to nerve function and memory. These findings suggest that even low doses of atrazine can negatively impact honey bee cognition, which may contribute to broader declines in bee populations. Understanding these effects is critical for developing safer agricultural practices and protecting pollinators, which are vital for food production and ecosystem stability. Our research highlights the need for stricter regulations on pesticide use to ensure the health of bees and the sustainability of global agriculture.

Abstract: Honey bees (*Apis mellifera*) are essential pollinators, responsible for the pollination of over 80% of crops and flowering plants globally. However, there is concern that the extensive use of pesticides, particularly atrazine, can harm pollinators. Despite the widespread use of atrazine, the sublethal effects on honey bees remain unclear. This study investigated the effects of atrazine on honey bee sucrose sensitivity and clarified the underlying molecular mechanisms using transcriptomic analyses. Atrazine exposure reduced the sucrose sensitivity of honey bees substantially, likely through the inhibition of functional genes associated with cognition in the brain. Genes related to neurodegenerative diseases and behavior were differentially expressed in response to atrazine. These findings provide novel insights into the neurophysiological and behavioral effects of atrazine on honey bees, contributing to a better understanding of pesticide risks and informing future environmental regulations.

Keywords: honey bees; atrazine; sucrose sensitivity; neurotoxicity; RNA sequencing

1. Introduction

Honey bees (*Apis mellifera*) are the most important pollinators in nature and are responsible for pollinating over 80% of crops and flowering plants globally [1]. Since the 20th century, the widespread use of pesticides has led to mass bee deaths and a drastic reduction in bee populations across many countries [2,3]. This has become a globally recognized issue in terms of environmental pollution and biological safety. Over the

https://doi.org/10.3390/insects16050491

past decade, the ban on high-toxicity pesticides has led to the increased use of low- and moderate-toxicity pesticides. Although deaths due to acute poisoning in bee populations have become less frequent, the sublethal effect of low-toxicity pesticides on bees has attracted increasing attention from researchers [4].

Atrazine, the second most widely used herbicide globally, has raised considerable concerns owing to its pervasive and improper application, resulting in severe environmental contamination and associated health risks [5]. It exhibits a broad spectrum of toxicological effects across various animal taxa, functioning as a neurotoxin, reproductive toxin, and endocrine disruptor [6]. There is increasing evidence that atrazine exerts detrimental effects on insect species. For instance, atrazine exposure disrupts immune system regulation in honey bees, thereby increasing their susceptibility to pathogenic infections [7]. In addition, experimental studies have demonstrated that atrazine exposure adversely affects longevity, prolongs developmental duration, and reduces body size in *Drosophila melanogaster* [8]. Female flies exposed to atrazine show a reduction in exploratory activity, an increase in immobility time, and dopamine system disruption [9]. Moreover, adult insects have been reported to exhibit a greater sensitivity to atrazine and selenium compared with that at larval stages, and the deleterious effects of these contaminants are more pronounced in terrestrial ecosystems than in aquatic environments [10]. Atrazine demonstrates significant neurotoxic properties in vertebrate models. Experimental studies in zebrafish *(Danio rerio)* have established that this herbicide induces behavioral deficits in both defensive and social responses through the specific inhibition of acetylcholinesterase activity in central nervous system tissues [11]. The hippocampus, a critical brain region responsible for learning and memory, is particularly susceptible to such effects in mice (*Mus musculus*) [12,13]. Exposure to atrazine during pregnancy and lactation in mice leads to increased apoptosis and dark neurons in the hippocampus of offspring, ultimately impairing their learning and spatial memory abilities [14]. Genovese et al. [15] found that atrazine exposure leads to behavioral changes and impaired motor memory in mice, with more pronounced effects in old mice than in young mice.

Research on the sublethal effects of atrazine on honey bees remains limited [5], despite the widespread use of this agent and the importance of honey bees for food production and ecosystem stability. In this study, we evaluated the impact of atrazine on honey bee's sucrose sensitivity and explored the underlying mechanisms through a transcriptomic analysis. By elucidating the effects of atrazine on honey bees at the molecular and behavioral levels, this research contributes to a deeper understanding of pesticide-induced stress in pollinators and provides a basis for conservation strategies and regulatory policies to mitigate the risks posed by herbicide exposure.

2. Materials and Methods

2.1. Honey Bees and Atrazine Preparation

Honey bees (*Apis mellifera*) were raised at Yangzhou University, China. Atrazine (99.17% TC; MedChemExpress, Princeton, NJ, USA) powder was initially dissolved in dimethyl sulfoxide, and 0.5% (3.73 mg/L) of the median lethal dose (LD_{50} = 746 mg/kg) for honey bees was used as the test dose [16]. Exposure to atrazine for six consecutive days did not lead to any observable increase in honey bee mortality under the tested conditions in our previous study [7].

2.2. Proboscis Extension Response (PER) Experiment

Three frames from different hives were placed in a dark incubator at 34 °C and 70% relative humidity. Newly emerged worker bees from different frames were divided into two groups, with three cup cages per group as biological replicates, each consisting of

20–30 bees. All bees were fed ample pollen bread and sucrose solution containing either 0 mg/L (CK) or 3.73 mg/L atrazine (AT) for 6 d [7].

Individual gustatory responses were measured using the classical proboscis extension response (PER) experiment [17]. The PER is a taste related behavior that is fundamental for olfactory discrimination and colony performance in honey bees. Six-day-old bees were tested for their proboscis extension response to sucrose solutions at concentrations of 0.1%, 0.3%, 1%, 3%, 10%, and 30%, following previously established protocols [18], with some modifications. Before the test, the bees were starved for 3–4 h in an incubator by removing the sucrose solution and bee bread from the cup cage. All bees were anesthetized by chilling and fixed in small tubes with only their heads moving. Before the presentation of each sucrose solution, bees that reacted to pure water or did not react to high concentrations of sucrose solution were excluded. Honey bees will reflexively extend the proboscis when the antennae are stimulated by a sufficiently concentrated sucrose solution. To weaken the effect of physical stress, the stimulation interval between water and sucrose solution was 5 min. The response was arbitrarily quantified with scores of 0–6, with 1 representing a response only to the highest sucrose concentration and 6 representing a response to all concentrations presented (Table S1). After the PER experiment, all bee brains were dissected and stored in a −80 °C ultra-low-temperature environment.

2.3. Brain Gene Expression Analysis

Gene expression profiles in the brains of the bees were evaluated. Each group contained three biological replicates. RNA was extracted from honey bee brains according to the manufacturer's recommendations (Vazyme, Nanjing, China) and sequenced using the Illumina NovaSeq 6000 Sequencing System. Raw reads containing adapters or low-quality bases were filtered using fastp 0.18.0. Clean reads were mapped to the *A. mellifera* reference genome, HAv3.1 (GenBank accession number GCA_003254395.2), using HISAT2 2.1.0. Subsequently, a gene count table was obtained using StringTie, and differentially expressed genes (DEGs) were identified using DESeq2 (version 1.32.0). Genes with adjusted $p < 0.05$ and |fold-change| ≥ 1.5 were considered significantly differentially expressed.

A hierarchical clustering analysis of DEGs was performed to examine global expression patterns. Gene Ontology (GO) enrichment analysis was subsequently conducted to classify DEGs into functionally relevant categories, including biological processes, molecular functions, and cellular components. Additionally, Kyoto Encyclopedia of Genes and Genomes (KEGG) pathway enrichment analysis was carried out using a hypergeometric distribution model to identify significantly overrepresented metabolic and signaling pathways. Functional annotation and pathway mapping of DEGs were conducted based on the KEGG database to gain insights into the underlying biological mechanisms.

To validate the differential gene expression identified through RNA sequencing, quantitative reverse transcription PCR (qRT-PCR) was performed using a QuantStudio™ 3 Real-Time PCR System (Thermo Fisher Scientific, Waltham, MA, USA). Total RNA was extracted as previously described [7] and subsequently reverse-transcribed into complementary DNA using a commercial kit (Vazyme, Nanjing, China). The synthesized cDNA was then used as a template for quantitative analysis of selected genes following the manufacturer's qPCR protocol (Vazyme, Nanjing, China). Relative gene expression levels between control and atrazine-treated groups were calculated using the $2^{-\Delta\Delta CT}$ method, with *Apis mellifera* actin employed as the endogenous reference gene. Gene-specific primers were designed using NCBI Primer-BLAST and are listed in Supplementary Table S2.

2.4. Statistical Analysis

Sucrose sensitivity scores for the two groups of bees were analyzed using Mann–Whitney U tests (* $p < 0.05$; ** $p < 0.01$; *** $p < 0.001$) implemented in SPSS 27.0 software. Bioinformatic analyses of gene expression and visualization were performed using R software.

3. Results

3.1. Atrazine Reduces Sucrose Sensitivity in Honey Bees

The gustatory response scores for honey bees exposed to atrazine were significantly lower than those in the control group ($p < 0.01$); the scores in the atrazine group were 70.92% of those in the control group (Figure 1). These results suggested that atrazine exposure impairs the sucrose sensitivity of bees.

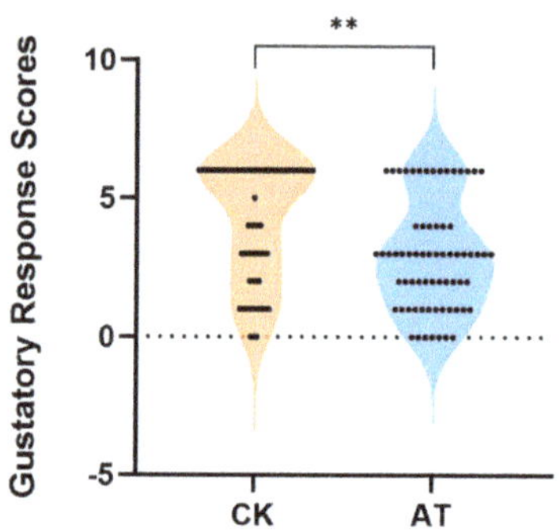

Figure 1. Comparison of sucrose sensitivity with and without atrazine treatment in the honey bee. CK: normal honey bee ($n = 68$); AT: atrazine-treated honey bee ($n = 71$); atrazine reduced honey bee sensitivity to sucrose significantly. Mann Whitney U test, ** $p < 0.01$.

3.2. Atrazine Changes the Functional Gene Expression in Honey Bee Brain

The differential gene expression patterns following the atrazine treatment were analyzed to elucidate the molecular mechanisms underlying the altered olfactory sensitivities in honey bees. In total, 10,448 DEGs were identified (Table S3). Among these, 112 genes were upregulated and 72 genes were downregulated in the brains of atrazine-treated honey bees (Figure 2a; Table S3), indicating a significant shift in the gene expression associated with atrazine exposure. The control group showed less variation in gene expression than that in atrazine-treated bees (Figure 2b), suggesting variability in the gene expression responses among individuals exposed to the chemical stressor.

Detoxification pathways play a crucial role in the honey bee response to xenobiotic stress, and our findings indicate that these pathways were activated in response to atrazine exposure. Specifically, detoxification related genes, such as cytochrome P450 9e2, 4C1, and glutathione S-transferase D1, were significantly upregulated, which is consistent with the activation of the detoxification system. However, the expression of several functional genes, particularly those related to neuronal function, was suppressed in honey bees exposed to atrazine. For example, Neprilysin-2, FoxP, ataxin-2, and the neuropeptide Y receptor were downregulated.

To gain insights into the functional implications of the DEGs, a GO enrichment analysis focusing on the molecular function domain, along with a KEGG pathway enrichment analysis, was performed. In the molecular function category of the GO analysis, DEGs were primarily enriched in terms such as the binding and catalytic activity, followed by the transporter activity and structural molecule activity (Figure S1). These results suggest that the identified DEGs are predominantly involved in molecular interactions, enzymatic activities, and transport related functions. The KEGG pathway enrichment analysis further revealed several significantly enriched pathways (Figure 3). Among these, protein digestion and absorption, and the Toll and Imd signaling pathway were the most prominent, indicating potential disruptions in nutrient utilization and innate immune responses. Notably, pathway associated with multiple neurodegenerative disease was enriched. These results suggest that atrazine exposure may interfere with neural development, which could contribute to the observed reduction in sucrose sensitivity.

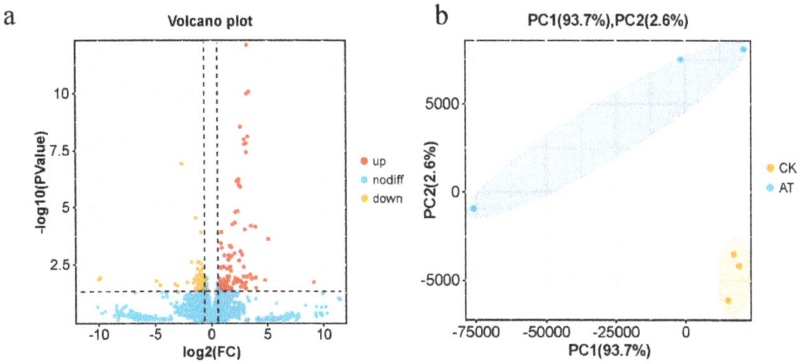

Figure 2. Transcriptome profiles of the brains of atrazine-treated and normal honey bees. (**a**) In total, 10,448 genes were detected through RNA sequencing, including 112 significantly upregulated (red) and 72 downregulated (yellow) genes in atrazine-treated honey bee brains compared with normal honey bees. (**b**) Principal coordinate analysis of the gene expression profiles. Samples in the control group were clustered, whereas samples in the atrazine-treated group were more dispersed.

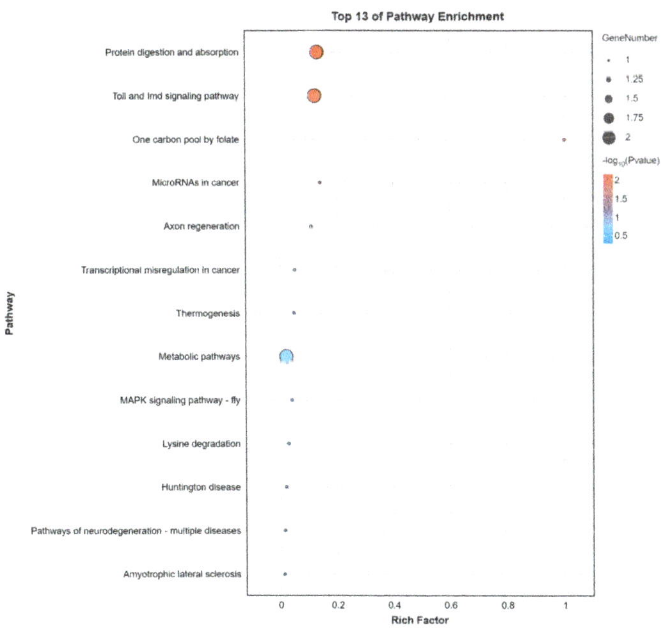

Figure 3. Top 13 significantly enriched Kyoto Encyclopedia of Genes and Genomes (KEGG) pathways associated with differentially expressed genes (DEGs). The *x*-axis indicates the KEGG enrichment score and the *y*-axis indicates the pathway names. Rich factor is a measure of the enrichment of a given pathway, calculated as the ratio of the number of DEGs in a pathway to the total number of genes involved in that pathway.

The genes selected for validation were primarily involved in neural function. The RT-qPCR analysis confirmed their significant downregulation in atrazine-treated bees, in agreement with the RNA-seq results (Figure S2). This transcriptional repression suggests that atrazine exposure may disrupt neuronal processes, potentially contributing to the behavioral deficits observed.

4. Discussion

Atrazine is the second most widely used pesticide in commercial agriculture worldwide [19]. However, its effects on honey bee neurons have not been well studied. The current study aimed to evaluate the sublethal effects of atrazine on honey bee sucrose sensitivity and elucidate the potential underlying molecular mechanisms using a transcriptomic approach. The results provide strong evidence that atrazine exposure alters the sensory and behavioral responses of honey bees substantially, specifically their sucrose sensitivity.

Olfaction plays a crucial role in various aspects of the honey bee ecology, including social communication, task division, queen rearing, individual development, swarming, and reproduction [20]. Honey bees demonstrate flower constancy because of their advanced sensory abilities and efficient capacity to learn and remember specific odors, which increases the ability of a colony to forage in a precise and efficient manner [21].

Previous studies have shown that exposure to pesticides, such as glyphosate, can alter honey bee preferences for certain food sources, which may further disrupt the foraging efficiency and colony health. These changes in preference, combined with a reduced sucrose sensitivity, could have significant implications for honey bee behavior and their role in pollination [22,23]. In this study, atrazine-treated honey bees exhibited a marked reduction in sucrose sensitivity, as measured by the PER, suggesting that atrazine impairs the ability of bees to effectively detect and respond to feed sources. This, in turn, could directly affect their foraging behavior and, consequently, their role in pollination.

Detoxification pathways play important roles in the ability of honey bees to cope with xenobiotic stress [24]. In general, the toxin structure is enzymatically altered primarily by the cytochrome P450 monooxygenase gene family and then solubilized and transported by detoxification enzymes, such as glutathione *S*-transferases [24]. The expression levels of some genes related to detoxification, such as cytochrome P450 9e2, 4C1, and glutathione *S*-transferase D1, were elevated, suggesting that the detoxification system of bees was activated. Glutathione *S*-transferase may be upregulated to reduce atrazine-induced oxidative stress [25,26]. Genes related to the Toll and Imd signaling pathways, which are crucial for immune regulation, were also upregulated in the atrazine-treated honey bees. This suggests that atrazine exposure activates immune related processes, potentially affecting the ability to respond to environmental stressors. Low-dose atrazine exposure induces DNA damage and activates glutathione S-transferase activity in zebrafish. However, this enzymatic activity shows a concentration-dependent inhibition at higher atrazine concentrations, suggesting the compromised detoxification capacity of the antioxidant defense system [27].

Various functional genes were suppressed when honey bees were exposed to atrazine, including neuron related genes. Neprilysin-2, a protease that degrades the amyloid-β peptide [27], has been shown to play an important role in protecting against Alzheimer's disease [28–32] and is positively associated with cognitive function in humans. Although these findings are based on mammalian studies, they provide valuable insights that may offer a reference point for understanding the potential neural functions of Neprilysin-2 in honey bees, despite the phylogenetic distance between species. *Drosophila* FoxP plays an important role in α-lobe mushroom body formation and is believed to be involved in neurodevelopmental processes and behaviors; individuals with mutant FoxP show mild cognitive impairment [33,34]. In addition to Neprilysin-2 and FoxP, ataxin-2-, neuropeptide Y receptor-, and neurofilament heavy polypeptide-like proteins function in various neurodegenerative diseases [35,36]. Moreover, KEGG pathway analysis indicated that several DEGs in honey bee brains were associated with neurodegenerative disease related signaling pathways (Figure 3). The observed downregulation of these functional genes suggests a potential disruption of neural processes following atrazine exposure, which may underlie the reduced sucrose sensitivity observed in treated bees.

5. Conclusions

In conclusion, this study provides preliminary evidence that sublethal atrazine exposure can modulate sucrose sensitivity and alter the brain gene expression in honey bees under laboratory conditions. These findings highlight the potential risk that even low-dose pesticide exposure may pose to pollinator health. However, since the current study was conducted under controlled laboratory conditions, further comprehensive evaluations under field conditions are necessary to fully assess the ecological relevance and broader impacts of atrazine exposure.

Supplementary Materials: The following supporting information can be downloaded at: https://www.mdpi.com/article/10.3390/insects16050491/s1, Figure S1: GO enrichment analysis of DEGs in the molecular function category; Figure S2: Confirmation of DEGs by RT-qPCR. Table S1: Scoring Methods of PER; Table S2: Primers used in this study; Table S3: Gene expression analysis.

Author Contributions: X.H.: conceptualization, methodology, software, data curation, writing—original draft. Z.X., G.M., J.X., Y.P. and M.C.: conceptualization, methodology, writing—original draft. Z.L., T.J. and K.W.: project administration, funding acquisition and supervision. K.W.: writing—review & editing. All authors have read and agreed to the published version of the manuscript.

Funding: Financial support was granted by the Natural Science Foundation of Yangzhou (K.W., YZ202416), National Natural Science Foundation of China (3240200019 and 32272935), and Earmarked Fund for Modern Agro-industry Technology Research System (CARS-44, T.J.).

Data Availability Statement: Raw sequence reads have been deposited in the NCBI SRA database under the BioProject accession no. PRJNA783076.

Conflicts of Interest: The authors declare no conflicts of interest.

Abbreviations

The following abbreviations are used in this manuscript:

PER	Proboscis extension response
DEG	Differentially expressed gene
KEGG	Kyoto Encyclopedia of Genes and Genomes

References

1. Hung, K.L.J.; Kingston, J.M.; Albrecht, M.; Holway, D.A.; Kohn, J.R. The worldwide importance of honey bees as pollinators in natural habitats. *Proc. Roy. Soc. B* **2018**, *285*, 20172140. [CrossRef] [PubMed]
2. Pervez, M.; Manzoor, F. Honey bee losses and pesticides threat: An Asian perspective. *J. Apic. Res.* **2023**, *62*, 64–75. [CrossRef]
3. Sgolastra, F.; Medrzycki, P.; Bortolotti, L.; Maini, S.; Porrini, C.; Simon-Delso, N.; Bosch, J. Bees and pesticide regulation: Lessons from the neonicotinoid experience. *Biol. Conserv.* **2020**, *241*, 108356. [CrossRef]
4. Barbosa, W.F.; Smagghe, G.; Guedes, R.N.C. Pesticides and reduced-risk insecticides, native bees and pantropical stingless bees: Pitfalls and perspectives. *Pest Manag. Sci.* **2015**, *71*, 1049–1053. [CrossRef]
5. Smith, A.E.; Muir, D.C.; Grover, R. The triazine herbicides. In *Anal of Pest in Water Anal Nitrogen Cont Pest*; CRC Press: Boca Raton, FL, USA, 2020; pp. 213–239.
6. Stradtman, S.C.; Freeman, J.L. Mechanisms of neurotoxicity associated with exposure to the herbicide atrazine. *Toxics* **2021**, *9*, 207. [CrossRef]

7. Wang, K.; Cai, M.; Sun, J.; Chen, H.; Lin, Z.; Wang, Z.; Niu, Q.; Ji, T. Atrazine exposure can dysregulate the immune system and increase the susceptibility against pathogens in honeybees in a dose-dependent manner. *J. Hazard. Mater.* **2023**, *452*, 131179. [CrossRef]
8. Marcus, S.R.; Fiumera, A.C. Atrazine exposure affects longevity, development time and body size in *Drosophila melanogaster*. *J. Insect Physiol.* **2016**, *91*, 18–25. [CrossRef] [PubMed]
9. Figueira, F.H.; de Quadros Oliveira, N.; de Aguiar, L.M.; Escarrone, A.L.; Primel, E.G.; Barros, D.M.; da Rosa, C.E. Exposure to atrazine alters behaviour and disrupts the dopaminergic system in *Drosophila melanogaster*. *Comp. Biochem. Physiol. Part C Toxicol. Pharmacol.* **2017**, *202*, 94–102. [CrossRef]
10. Henry, B.L.; Wesner, J.S. Severing ties: Different responses of larval and adult aquatic insects to atrazine and selenium. *Environ. Sci. Technol.* **2018**, *52*, 8848–8857. [CrossRef]
11. Schmidel, A.J.; Assmann, K.L.; Werlang, C.C.; Bertoncello, K.T.; Francescon, F.; Rambo, C.L.; Beltrame, G.M.; Calegari, D.; Batista, C.B.; Blaser, R.E. Subchronic atrazine exposure changes defensive behaviour profile and disrupts brain acetylcholinesterase activity of zebrafish. *Neurotoxicol. Teratol.* **2014**, *44*, 62–69. [CrossRef]
12. Khoshdel-Sarkarizi, H.; Hami, J.; Mohammadipour, A.; Sadr-Nabavi, A.; Mahmoudi, M.; Kheradmand, H.; Peyvandi, M.; Nourmohammadi, E.; Haghir, H. Developmental regulation and lateralization of GABA receptors in the rat hippocampus. *Int. J. Dev. Neurosci.* **2019**, *76*, 86–94. [CrossRef] [PubMed]
13. Baradaran, R.; Khoshdel-Sarkarizi, H.; Kargozar, S.; Hami, J.; Mohammadipour, A.; Sadr-Nabavi, A.; Peyvandi Karizbodagh, M.; Kheradmand, H.; Haghir, H. Developmental regulation and lateralisation of the α7 and α4 subunits of nicotinic acetylcholine receptors in developing rat hippocampus. *Int. J. Dev. Neurosci.* **2020**, *80*, 303–318. [CrossRef] [PubMed]
14. Rastegar-Moghaddam, S.H.; Mohammadipour, A.; Hosseini, M.; Bargi, R.; Ebrahimzadeh-Bideskan, A. Maternal exposure to atrazine induces the hippocampal cell apoptosis in mice offspring and impairs their learning and spatial memory. *Toxin Rev.* **2019**, *38*, 298–306. [CrossRef]
15. Genovese, T.; Siracusa, R.; Fusco, R.; D'Amico, R.; Impellizzeri, D.; Peritore, A.F.; Crupi, R.; Gugliandolo, E.; Morabito, R.; Cuzzocrea, S. Atrazine inhalation causes neuroinflammation, apoptosis and accelerating brain aging. *Int. J. Mol. Sci.* **2021**, *22*, 7938. [CrossRef] [PubMed]
16. Helmer, S.H.; Kerbaol, A.; Aras, P.; Jumarie, C.; Boily, M. Effects of realistic doses of atrazine, metolachlor, and glyphosate on lipid peroxidation and diet-derived antioxidants in caged honey bees (*Apis mellifera*). *Environ. Sci. Pollut. Res.* **2015**, *22*, 8010–8021. [CrossRef]
17. Ramírez, G.P.; Martínez, A.S.; Fernández, V.M.; Corti Bielsa, G.; Farina, W.M. The influence of gustatory and olfactory experiences on responsiveness to reward in the honeybee. *PLoS ONE* **2010**, *5*, e13498. [CrossRef]
18. Zhang, Z.; Mu, X.; Shi, Y.; Zheng, H. Distinct roles of honeybee gut bacteria on host metabolism and neurological processes. *Microbiol. Spectr.* **2022**, *10*, e02438-21. [CrossRef]
19. Centanni, M.; Ricci, G.; De Girolamo, A.; Romano, G.; Gentile, F. A review of modeling pesticides in freshwaters: Current status, progress achieved and desirable improvements. *Environ. Pollut.* **2023**, *316*, 120553. [CrossRef]
20. Caretta, M. Neurobiology of chemical communication. CRC Press: Boca Raton, FL, USA, 2014. [CrossRef]
21. Paoli, M.; Giurfa, M. Pesticides and pollinator brain: How do neonicotinoids affect the central nervous system of bees? *Eur. J. Neurosci.* **2024**, *60*, 5927–5948. [CrossRef]
22. Farina, W.M.; Balbuena, M.S.; Herbert, L.T.; Mengoni Goñalons, C.; Vázquez, D.E. Effects of the herbicide glyphosate on honey bee sensory and cognitive abilities: Individual impairments with implications for the hive. *Insects* **2019**, *10*, 354. [CrossRef]
23. Abuagla, M.I.; Iqbal, J.; Raweh, H.S.; Alqarni, A.S. Insight into olfactory learning, memory, and mortality of *Apis mellifera jemenitica* after exposure to acetamiprid insecticide. *Insects* **2024**, *15*, 473. [CrossRef]
24. Berenbaum, M.R.; Johnson, R.M. Xenobiotic detoxification pathways in honey bees. *Curr. Opin. Insect Sci.* **2015**, *10*, 51–58. [CrossRef]
25. Jin, Y.; Wang, L.; Chen, G.; Lin, X.; Miao, W.; Fu, Z. Exposure of mice to atrazine and its metabolite diaminochlorotriazine elicits oxidative stress and endocrine disruption. *Environ. Toxicol. Pharmacol.* **2014**, *37*, 782–790. [CrossRef]
26. Zhang, C.; Li, X.-N.; Xiang, L.-R.; Qin, L.; Lin, J.; Li, J.-L. Atrazine triggers hepatic oxidative stress and apoptosis in quails (*Coturnix C. coturnix*) via blocking Nrf2-mediated defense response. *Ecotoxicol. Environ. Saf.* **2017**, *137*, 49–56. [CrossRef] [PubMed]
27. Zhu, L.; Dong, X.; Xie, H.; Wang, J.; Wang, J.; Su, J.; Yu, C. DNA damage and effects on glutathione S-transferase activity induced by atrazine exposure in zebrafish (*Danio rerio*). *Environ. Toxicol.* **2011**, *26*, 480–488. [CrossRef] [PubMed]
28. Hafez, D.; Huang, J.Y.; Huynh, A.M.; Valtierra, S.; Rockenstein, E.; Bruno, A.M.; Lu, B.; DesGroseillers, L.; Masliah, E.; Marr, R.A. Neprilysin-2 is an important β-amyloid degrading enzyme. *Am. J. Pathol.* **2011**, *178*, 306–312. [CrossRef]
29. Huang, S.-M.; Mouri, A.; Kokubo, H.; Nakajima, R.; Suemoto, T.; Higuchi, M.; Staufenbiel, M.; Noda, Y.; Yamaguchi, H.; Nabeshima, T. Neprilysin-sensitive synapse-associated amyloid-β peptide oligomers impair neuronal plasticity and cognitive function. *J. Biol. Chem.* **2006**, *281*, 17941–17951. [CrossRef] [PubMed]

30. Spencer, B.; Marr, R.A.; Rockenstein, E.; Crews, L.; Adame, A.; Potkar, R.; Patrick, C.; Gage, F.H.; Verma, I.M.; Masliah, E. Long-term neprilysin gene transfer is associated with reduced levels of intracellular Abeta and behavioral improvement in APP transgenic mice. *BMC Neurosci.* **2008**, *9*, 109. [CrossRef]
31. Liu, Y.; Studzinski, C.; Beckett, T.; Guan, H.; Hersh, M.A.; Murphy, M.P.; Klein, R.; Hersh, L.B. Expression of neprilysin in skeletal muscle reduces amyloid burden in a transgenic mouse model of Alzheimer disease. *Mol. Ther.* **2009**, *17*, 1381–1386. [CrossRef]
32. Huang, J.Y.; Hafez, D.M.; James, B.D.; Bennett, D.A.; Marr, R.A. Altered NEP2 expression and activity in mild cognitive impairment and Alzheimer's disease. *J. Alzheimer's Dis.* **2012**, *28*, 433–441. [CrossRef]
33. Castells-Nobau, A.; Eidhof, I.; Fenckova, M.; Brenman-Suttner, D.B.; Scheffer-de Gooyert, J.M.; Christine, S.; Schellevis, R.L.; Van der Laan, K.; Quentin, C.; Van Ninhuijs, L. Conserved regulation of neurodevelopmental processes and behavior by FoxP in *Drosophila*. *PLoS ONE* **2019**, *14*, e0211652. [CrossRef] [PubMed]
34. Reuter, M.S.; Riess, A.; Moog, U.; Briggs, T.A.; Chandler, K.E.; Rauch, A.; Stampfer, M.; Steindl, K.; Gläser, D.; Joset, P. FOXP2 variants in 14 individuals with developmental speech and language disorders broaden the mutational and clinical spectrum. *J. Med. Genet.* **2017**, *54*, 64–72. [CrossRef] [PubMed]
35. Ostrowski, L.A.; Hall, A.C.; Mekhail, K. Ataxin-2: From RNA control to human health and disease. *Genes* **2017**, *8*, 157. [CrossRef] [PubMed]
36. Broberger, C.; Visser, T.J.; Kuhar, M.J.; Hökfelt, T. Neuropeptide Y innervation and neuropeptide-Y-Y1-receptor-expressing neurons in the paraventricular hypothalamic nucleus of the mouse. *Neuroendocrinology* **1999**, *70*, 295–305. [CrossRef]

MDPI AG
Grosspeteranlage 5
4052 Basel
Switzerland
Tel.: +41 61 683 77 34

Insects Editorial Office
E-mail: insects@mdpi.com
www.mdpi.com/journal/insects